Advances in Cement and Concrete

3rd International Conference on Advances in Cement and Concrete (ICACCR-2024), Ladoke Akintola University of Technology, Ogbomoso, Nigeria, 5-8 November 2024

Editors
Akeem Ayinde Raheem[1], Bolanle D. Ikotun[2]

[1]Ladoke Akintola University of Technology, Ogbomoso, Nigeria
[2]University of South Africa, South Africa

Peer review statement

All papers published in this volume of "Materials Research Proceedings" have been peer reviewed. The process of peer review was initiated and overseen by the above proceedings editors. All reviews were conducted by expert referees in accordance to Materials Research Forum LLC high standards.

Published under License by **Materials Research Forum LLC**
Millersville, PA 17551, USA

Published as part of the proceedings series
Materials Research Proceedings
Volume 51 (2025)

ISSN 2474-3941 (Print)
ISSN 2474-395X (Online)

ISBN 978-1-64490-352-0 (Print)
ISBN 978-1-64490-353-7 (eBook)

This book contains information obtained from authentic and highly regarded sources. Reasonable efforts have been made to publish reliable data and information, but the author and publisher cannot assume responsibility for the validity of all materials or the consequences of their use. The authors and publishers have attempted to trace the copyright holders of all material reproduced in this publication and apologize to copyright holders if permission to publish in this form has not been obtained. If any copyright material has not been acknowledged please write and let us know so we may rectify in any future reprint.

Distributed worldwide by

Materials Research Forum LLC
105 Springdale Lane
Millersville, PA 17551
USA
https://mrforum.com

Manufactured in the United State of America
10 9 8 7 6 5 4 3 2 1

Table of Contents

Preface

ICACCR 2024

The third edition of the International Conference on Advances in Cement and Concrete Research (ICACCR 2024) held between November 5-8, 2024, in Ladoke Akintola University of Technology, Ogbomoso, Nigeria, continued as a series of events marked towards the development of innovative and sustainable construction materials for nation building. The conference themed **"Applications of Cement and Concrete Materials for Smart, Resilient, and Sustainable Infrastructures "**covered the development of green building materials, synthesis and applications of nanoparticles, self healing and self sensing cement composites and High-Performance Concrete (HPC). This event provided a platform for researchers, contractors, academia and industries to deliberate on the challenges and potential way forward as well as new frontier in the production, processing, analysis and testing of cement and concrete materials for sustainable development.

In addition, the conference gave opportunities for scientific and technical exchange of knowledge and further strengthening of collaborative efforts in Concrete Technology. The event captured participants around the world who presented different topics trending in the field of Concrete Technology thus given students an enabling atmosphere to present their research findings. The articles presented in this proceeding had undergone a blind double peer-review process by the international technical scientific committee.

We would like to express our sincere gratitude to all the organising committees and participants for their contributions towards the success of the conference.

Committees

Organising Committee

Prof. Akeem A. Raheem - LAUTECH, Nigeria (Conference Chairman)
Prof. Bolanle D. Ikotun - UNISA, South Africa (Co-Conference Chairman)
Prof. Johnson R. Oluremi - LAUTECH, Nigeria (Conference Secretary Chairman)
Prof. Solomon O. Ajamu - LAUTECH, Nigeria (Fund Raising Chairman)
Dr. Solomon I. Adedokun - University of Lagos, Nigeria (Member)
Dr. Gideon O. Bamigboye - Covenant University (Member)
Dr. Blessing O. Orogbade - Olabisi Onabanjo University, Nigeria (Member)
Dr. Solomon Olakunle OYEBISI - Covenant University (Member)
Engr. Mukaila A. Anifowose - Federal Polytechnic Offa, Nigeria (Member)
Dr. Kazeem Ishola - Osun State University, Nigeria (Member)
Engr. A. O. Zimbili - University of South Africa (Member)
Dr Thabo Falayi - Namibia University of Science Technology (Member)
Dr. Abideen Ganiyu - Military Technological College, Muscat, Oman (Member)
Dr. E. O. Ibiwoye - Kwara State Polytechnic, Ilorin (Member)
Dr. Oluwaleke A. Olowu - Yaba College of Technology, Nigeria (Member)
Dr. Rasheed Abdulwahab - Kwara State University, Nigeria (Member)
Dr. Mutiu O. Kareem - Osun State University, Nigeria (Member)
Engr. Oladayo Awodele - LAUTECH, Nigeria (Member)
Mr. Samuel Akinloye - LAUTECH, Nigeria (Member)

Advances in Cement and Concrete
Materials Research Proceedings 51 (2025) 1-9

Materials Research Forum LLC
https://doi.org/10.21741/9781644903537-1

The impact of wood ash and nanosilica on the consistency, setting time, and flowability of Portland cement

Abiodun AKINWALE[1,a,*], Hussein WALIED[1,b], Akeem Ayinde RAHEEM[2,c]

[1]Department of Civil Engineering, University of South Africa Florida Campus, Johannesburg, South Africa

[2]Department of Civil Engineering, Ladoke Akintola University of Technology, Ogbomoso, Oyo State, Nigeria

[a]akinaae@unisa.ac.za; [b]hussiwam@unisa.ac.za; [c]aaraheem@lautech.edu.ng

Keywords: Woodash, Nanosilica, Consistency, Setting Time, Flowability, Cement Paste

Abstract. The building and construction sectors are sourcing innovative cementitious materials to enhance the conventional binder's performance, durability, and sustainability. Optimizing resources and developing cost-effective, sustainable materials over their lifecycle is essential. This study investigates the impact of wood ash (WA) and nanosilica (NS) on Portland cement's standard consistency, setting time, and flowability. The wood ash was sourced from Caterer's Services in Johannesburg, ground to finer particles, and sieved. Preliminary analyses were conducted on the ash, which included oxide composition, specific gravity, and particle size distribution. The control sample (CC) contained no wood ash or nanosilica, while the experimental groups contained varying amounts of wood ash (5% to 25%) and nanosilica (0.6%, 1.1%, and 1.7%). The investigation shows the consistency of the cement paste was enhanced at 15% WA, and the incorporation of 0,6% NS gave the best consistency compared to 1,1 and 1,7 % NS. Higher WA content, particularly at 25% replacement (WA25NS0), resulted in the longest setting time, reaching approximately 600 minutes, while the flowability increases as the WA percentage increases. The results demonstrated consistent performance in binders containing 15 % WA and the incorporation of 0,6 % NS compared to 1,1 and 1 7 % NS. Higher WA content, at 25 % replacement (WA25NS0), resulted in the longest setting time. Flowability increases as the WA percentage increases, peaking at WA15NS0, representing 33,12% compared to the control sample.

1. Introduction

The building and construction sectors are actively searching for innovative, cutting-edge cementitious materials. Enhanced cementitious system performance can be attained by resource optimization, developing more durable and environmentally friendly materials, and life cycle cost-effectiveness. The cement production industry has been an important one in the development of infrastructure constructed with cement composite concrete since the dawn of the twentieth century, including dams, dykes, and bricks, blocks [1]. Cement manufacturers must plan to decarbonize their industry to build climate-resilient buildings and concrete infrastructure. Consequently, researchers have been exploring alternative cementitious materials, including natural pozzolan, and industrial wastes, such as fly ash and coal bottom ash, to reduce atmospheric pollution. These materials contain silica and alumina, with a unique chemical activity known as "pozzolans [1]. They have various benefits, including greater ecological commitment and natural resource savings.

There is a lack of sufficient scientific data in the cement industry about wood ash (WA) as a supplementary cementing material, and it is necessary to explore its applications in the production of cement composites [2]. The addition of WA can affect the properties of cement pastes and mortars. Their study on the rheology, hydration, and microstructure of Portland cement pastes produced with ground Açai fiber found that 5% Açai fiber delayed cement hydration by around

Advances in Cement and Concrete   Materials Research Forum LLC
Materials Research Proceedings 51 (2025) 1-9   https://doi.org/10.21741/9781644903537-1

2.5 days. Several researchers, including Gopinath et al. [3], have focused on using sawdust ash as a replacement in mortars and concrete. Their work has significantly contributed to a better understanding of the durability and behaviour of mortar and concrete with wood ash. Additionally, Darwesh et al.) [4] researched mortar with wood waste ash. Their work showed a change in the flow properties of cement-ash pastes due to increased water requirement and an optimal replacement of 15% at 28 days of curing.

Freshly mixed concrete reacts with nanosilica particles to change its consistency, workability, and other characteristics [5]. Using nanosilica reduces the latent phase of setting time by around 1-2%, but the actual setting time would still be between 90 and 100 minutes [6]. Several studies asserted that the workability of concrete was impacted by nanosilica. The amount of water needed in concrete varied due to the nano-tiny silica particle size [7]. More fineness allows a material to absorb more water while mixing since it has a larger surface area [8]. A portion of the combined water, or water molecules, pushes towards the surface of the nanosilica particles due to their large surface area and unsaturated bonds, creating silanol (Si-OH) groups [9]. According to [10], the maximum amount of nano-silica that may be utilized for an acceptable range of slump flow is 3% by weight.

However, according to Jalal et al. [11]. HPSCC analysis, adding 2% nanosilica did not affect the workability of concrete, which was later confirmed by [10]. The presence of free water among the ultra-fine particles increased the rolling effects between the particles when nanosilica was appropriately dispersed and de-agglomerated, resulting in a notable increase of around 35% in the workability of concrete. The results suggest that a more consistent dispersion of nanoparticles can enhance concrete utilization. In this study, an in-depth characterization of the materials used was carried out. Consequently, this work aims to examine the Impact of wood ash and nanosilica on the consistency, setting time, and flowability of Portland Cement.

2. Materials and methods

2.1 Materials
Ordinary Portland cement class CEM I 52.5N conforming to SANS 50197-1 [12] was used in this investigation. The wood ash was sourced from a local canteen in Johannesburg, South Africa. The chemical and physical properties of the wood ash used are provided in Table 1. Nanosilica (NS) powder containing 99.9% particle SiO_2 with an average diameter of $\leq$150 μm and pore size of 6 nm. Its carbon content is 6,1%, while its Nitrogen content is 1,8%. It was purchased from Sigma-Adrich. Tap water was used in all mixtures.

2.2 Methods
2.2.1 Cement paste design and preparation.
This work used conventional cement paste designs. Wood ash (WA) was used as a replacement for cement in varying content of 5, 10, 15, 20, and 25 %, respectively. Water amount was determined as a prerequisite to achieving a standard consistency conforming to SANS 50196-3 [13] for preparing paste mixes. The control sample contains 100% Portland Cement (PC) without wood ash and Nanosilica. The experimental group was divided into four stages, each containing five mixes. The first group (WA05NS0 – WA25NS0) contains 5 to 25 % wood ash without nanosilica. The second group (WA05NS0,6 – WA25NS0,6) contains wood ash between 5 to 25 % with 0,6% nanosilica, while the third (WA05NS1,1 – WA25NS1,1) and fourth group (WA05NS1,7 – WA25NS1,7) contains wood ash between 5 to 25% wood ash containing 1,1 and 1,7 % nanosilica respectively. All mixes had a water-to-binder (w/b) ratio of 0.3 by mass. The cement and water were combined in an automatic mixer referencing standard practice for mechanical mixing of hydraulic cement pastes and mortars of plastic consistency [14] and stirred to produce the cement paste. The mixer, with a volume of 4.5 L capacity, was used to mix 500g of paste per batch to ensure sufficient material for testing. Initially, water, nanosilica-prepared

solution, cement, and wood ash were added to the mixing container, and the mixture was stirred at a low speed of 140 r/min for 1.5 minutes. After this initial period, the mixer was paused, and the walls of the mixing container were scraped to ensure uniformity. The mixer was then switched to a higher speed of 285 r/min for 1.5 minutes to achieve a homogenous paste.

2.2.3. Wood ash characterization

The physical and chemical properties of the wood ash are presented in Table 1. The densities, specific gravity, and fineness modulus were determined following standards [15,16]]. The wood ash sample in Table 1 contains various elemental oxides. The combined percentages of SiO_2, Al_2O_3, and Fe_2O_3 in the wood ash were greater than 70%, which correlates well with the study by [17]. This suggests that the wood ash is a suitable pozzolanic material that meets the requirements of [18].

Table 1, Chemical and physical properties of Wood ash

Wt.[%]	[%]	Wt.[%]	[%]	Wt.[%]	[%]
SiO_2	53.57	Al_2O_3	33.98	CaO	3.24
MgO	6.23	MnO	1.66	Na2O	-
K_2O	0.20	TiO_2	0.20	P_2O_5	0.33
SO_3	1.58	ZnO	2.66	SrO_2	0.08
ZrO_2	0.96	Cl	0.07	Fe_2O_3	6.12

Physical properties

Specific gravity	Mean Size [μm]	Density [kg/m³]
1.92	0.4	834

Table 2, Summary of formulations (replacement of cement)

Paste type	Cement [g)]	Ash content [%]	Wood ash [g]	Nanosilica [%]	Water [g]	W/B
CC	500	0	0	0	155	0.3
WA05NS0	475	5	17,857	0	175	0.3
WA10NS0	450	10	35,714	0	188	0.3
WA15NS0	425	15	53,571	0	190	0.3
WA20NS0	400	20	71,429	0	180	0.3
WA25NS0	375	25	89,286	0	180	0.3
WA05NS0,6	475	5	17,857	0.6	157	0.3
WA10NS0,6	450	10	35,714	0.6	152	0.3
WA15NS0,6	425	15	53,571	0.6	150	0.3
WA20NS0,6	400	20	71,429	0.6	149	0.3
WA25NS0,6	375	25	89,286	0.6	148	0.3
WA05NS1,1	475	5	17,857	1.1	156	0.3
WA10NS1,1	450	10	35,714	1.1	151	0.3
WA15NS1,1	425	15	53,571	1.1	149	0.3
WA20NS1,1	400	20	71,429	1.1	149	0.3
WA25NS1,1	375	25	89,286	1.1	148	0.3
WA05NS1,7	475	5	17,857	1.7	155	0.3
WA10NS1,7	450	10	35,714	1.7	150	0.3
WA15NS1,7	425	15	53,571	1.7	149	0.3
WA20NS1,7	400	20	71,429	1.7	148	0.3
WA25NS1,7	375	25	89,286	1.7	149	0.3

Advances in Cement and Concrete   Materials Research Forum LLC
Materials Research Proceedings 51 (2025) 1-9   https://doi.org/10.21741/9781644903537-1

3. Results and discussions

3.1 Consistency of cement paste

Figure 1(a-d) shows the consistency of wood ash cement paste without nanosilica compared to wood ash incorporating nanosilica. The increasing content of WA decreased the consistency of paste up to 15% WA and increased afterward. Incorporating 0,6% NS increased the consistency of WA cement paste up to 10% WA. However, the addition of 1,1 and 1,7% NS gave haphazard consistency. The behavioural pattern of increasing content of NS may be attributed to the unstable nature of atoms therein. This agrees with what was obtained elsewhere (Hailong Sun et al.) [19]

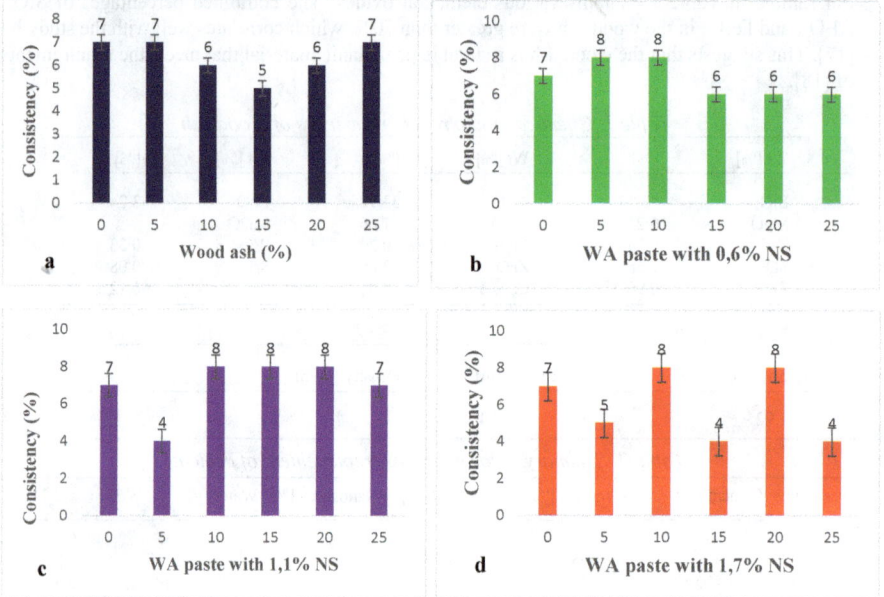

Fig. 1, Standard consistency of WA paste incorporating different % of NS

3.2 Setting time

Figure 2(a-d) displays each group's initial setting time results. The (WA05NS0 – WA25NS0) group shows an upward trend in the initial setting time as the percentage of WA increases, picking up at WA20NS0 (about 400 minutes). This suggests that the higher WA replacement without NS extends the initial setting time. Conversely, the (WA05NS0,6 – WA25NS0,6) group shows a drastic reduction in the initial setting time compared to the (WA05NS0 – WA25NS0) group, especially at WA05NS0,6 and WA15NS0,6, respectively, where the initial setting time is the lowest around 150 minutes. The introduction of the NS tends to accelerate the setting process, hence the reduction in the initial setting time. This performance is consistent with the work of Gopinath et al. [3], which suggests that nanosilica shortens the setting time of cement paste, reducing the fluidity and increasing the water demand. However, there is a sign of recovery in the setting time of the (WA05NS1,1 – WA25NS1,1) group after the initial drop experienced by the (WA05NS0,6 – WA25NS0,6) group.

Advances in Cement and Concrete
Materials Research Proceedings 51 (2025) 1-9

Materials Research Forum LLC
https://doi.org/10.21741/9781644903537-1

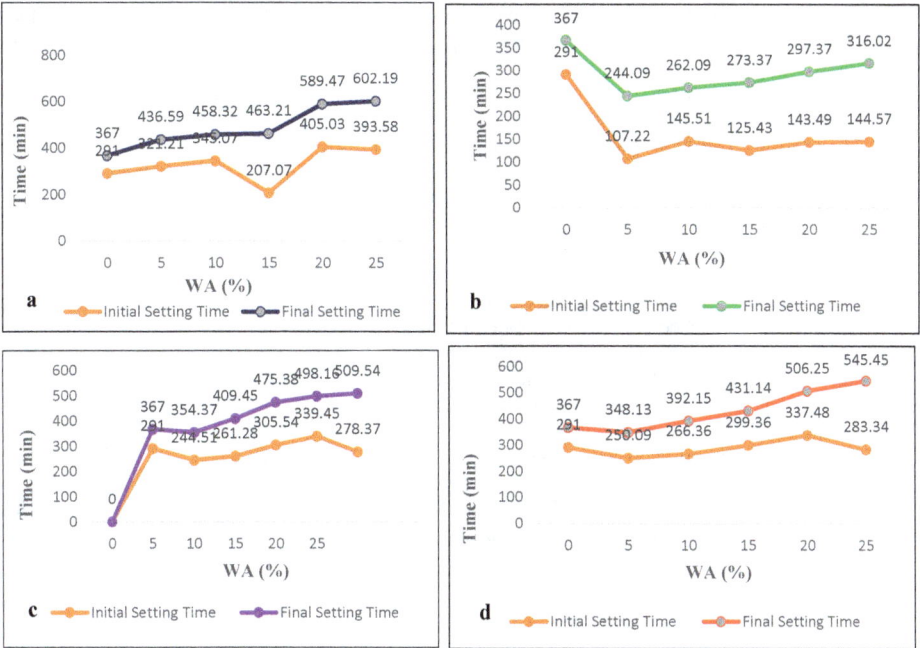

Fig. 2, Setting time of WA cement paste containing Nanosilica

Figure 2 (a-d) shows the results for the final setting time for the four groups compared to the control sample. The graph illustrates the cement paste's final setting time with varying cement replacement levels by combining WA and NS. The control specimen (CC) exhibits the lowest final setting time, around 300 minutes. As WA is introduced without nanosilica (WA0NS0 – WA25NS0), the setting time significantly increased, reaching a maximum of approximately 600 minutes for the sample with 25% WA replacement (WA25NS0). This trend suggests that wood ash alone delayed the hydration process of cement, likely due to its pozzolanic properties, which react slower than traditional cement hydration. The introduction of NS shows different behaviour. For instance, at lower WA replacement (WA05NS0,6 – WA25NS0,6), the addition of NS reduces the final setting time compared to the (WA05NS0 – WA25NS0) group. This indicates that Nanosilica enhances the reactivity of the cement mix, likely due to its high surface area and ability to promote early-stage hydration reactions. As the percentage of Nanosilica increases, the final setting time decreases for all WA replacement levels, but the effect becomes more pronounced in mixes with higher WA contents. (e.g., WA25NS1,7). The combination of WA and NS in moderate to higher quantities stabilizes the delayed hydration effect induced by WA without NS, resulting in a more manageable final setting time than WA-only mixes. The reduction observed in the initial setting time (IST) and final setting time (FST) as a result of the addition of nanosilica was consistent with the findings published by Pratibha et al. [20].

3.3 Flowability test

Paste mixes were characterized by flow behaviour, plasticity, and cohesion, whereas their workability defines consistency. Figure 6 shows the influence of nanosilica on the flowability of cement paste containing WA. It is clear from the result that the workability of the mixes without nanosilica, irrespective of the wood ash percentage inclusion (between 5 and 25%), is higher than the PC mix. The results indicate that the first group (WA05NS0-WA25NS0) containing wood ash between 5 to 25% without nanosilica gave the highest spread value compared to other groups containing nanosilica. An important finding is that nanosilica content affects workability characteristics in the mix corresponding to wood ash content, as illustrated in Fig. 3(a-d). The reduction in flowability occurs at certain WA+NS ratios such as WA10NS0,6 but experienced a slight recovery in WA+NS such as WA25NS1,7, though never reaching the initial high flowability compared to (WA05NS0-WA25NS0) group performance. The result also shows the combined effect of materials that have a significant effect on the paste performance compared to the control sample, such as cement and wood ash, cement and water; wood ash and water; wood ash and nanosilica on the flowability of the paste is significant. With the reduction in the flowability of the paste mixes containing various contents of nanosilica, the presence of nanosilica made cement paste thicker and accelerated the hydration process, as suggested by Qing et al. [21]. Pratibha et al. [20] believed that wood ash's spherical morphology helped increase cementitious materials' flowability. In contrast, nanosilica increases stiffness due to their higher specific surface area, thereby reducing the fluidity of wood ash cement-blended nanosilica paste with an increase in nanosilica [22]. Furthermore, it is observed that WA initially increased the flowability up to 15% WA and decreased afterward. Increasing the content of NS was observed to increase the flowability of the paste and this paste's flowability, which was evident with the addition of 1,7% NS. Although, the flow pattern observed in the case of 0,6 and 1,1 % NS was haphazard. This inconsistent pattern may be attributed to the unstable behaviour of NS, as reported by Singh et al. [23]

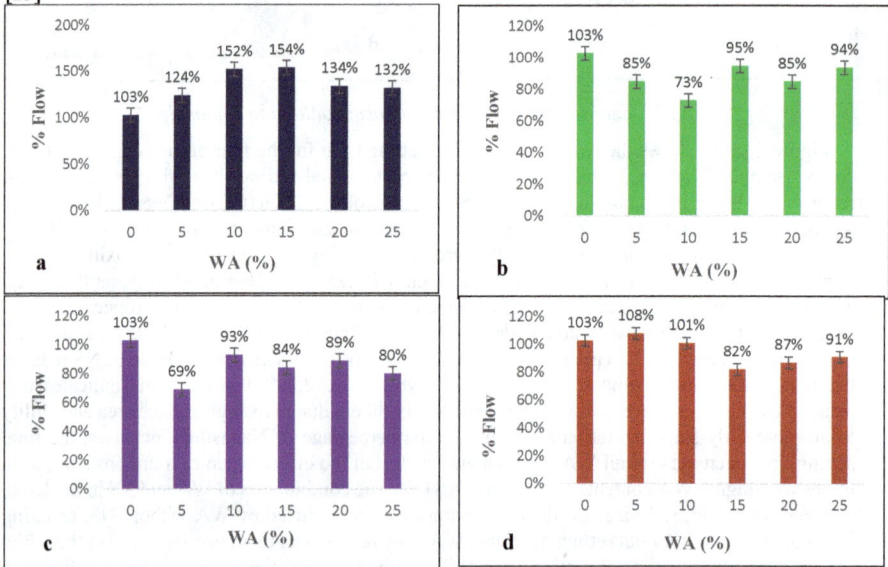

Fig. 3, Flowability of WA cement paste containing Nanosilica

Advances in Cement and Concrete
Materials Research Proceedings 51 (2025) 1-9

Materials Research Forum LLC
https://doi.org/10.21741/9781644903537-1

5. Conclusions and summary

This study focused on the combined effect of wood ash and nanosilica on cement paste's consistency, setting time, and flowability.

- The consistency of the cement paste was enhanced at 15% WA, and the incorporation of 0,6 % NS gave the best consistency as compared to 1,1 and 1 7 % NS.

- Higher WA content, particularly at 25% replacement (WA25NS0), resulted in the longest setting time, reaching approximately 600 minutes, indicating that WA slows down the hydration process of cement. The addition of NS progressively reduces the delay in the final setting time caused by WA, showing that NS counteracts the retardation effect of WA. As NS increases from (NS1,1 – NS1,7%), the setting time across all WA replacement levels decreases further, suggesting that higher amounts of NS accelerate the early hydration reactions.

- Flowability increases as WA percentage increases, peaking at WA15NS0. This suggests WA improves the flowability. This might be due to its particle structure or water-retaining properties. The addition of NS significantly reduces flowability, especially at higher percentages. This is a result of its absorbing water properties, making the mix denser and less workable.

Declarations

The authors declare that they have no conflicting interests or personal relationships that could have appeared to influence the work reported in this paper. Moreover, the authors would like to acknowledge the University of South Africa's contribution to this article's research, authorship, and/or publication.

References

[1] J.N. Wember, L.L.M. Elat, and C.D. Miyo, Impact of the partial substitution of cement and sand by ash from several types of wood species in cementitious materials manufacture volarisation in the industrial field. Discover Civil Engineering. (2024) https://doi.org/10.1007/s44290-024-00035-5

[2] J.M. Lessard, R. Gagné, and M.P. Jolin Rivard, Optimisation des cendres volantes et grossières de biomasse dans les bétons compactés au rouleau et dans les bétons moulés á sec, Mémoire de Maitise, Departement de Génie Civil, Université de Sherbrooke. (2016)

[3] K. Gopinath, K. Anuradha, R. Harisundar, and M. Saravanan. Utilization of sawdust in cement mortar & cement concrete. Int J Sci Eng Res. 6 (2015) 665-82.

[4] H.H.M. Darwesh, M.A. El-Suoud, Saw dust ash substitution for cement pastes – Part I Am J Constr. Build. Mater. 2 (2017) 1-9. https://doi.org/10.11648/j.ajasr.20170305.13

[5] N.V. Makarova, V.V. Potapov, A.V. Kozin, E.A. Chusovitin, A.V. Amosov, A.V. Nepomnyashiy. Influence of hydrothermal Nanosilica on mechanical properties of plain concrete. Key Eng. Mater. 744 744 KE, (2017)126-130. https://doi.rg/10.4028/www.scientific.net/KEM.

[6] S. Erdem, S. Hanbay, Z. Güler. Micromechanical damage analysis and engineering performance of concrete with colloidal nano-silica and demolished concrete aggregates. Constr. Build. Mater. 171 (2018) 634-642.
https://doi.org/10.1016/j.conbuildmat.2018.03.197.

[7] G. Li. Properties of high-volume fly ash concrete incorporating nano-SiO2, Cem. Concr. Res. 34 (2004) 1043-1049. https://doi.org/10.1016/j.cemconres.2003.11.013.

[8] L. Varghese, V.V.L. Kanta Rao, L. Parameswaran. Nanosilica-added concrete: strength and its correlation with time-dependent properties. Proc. Inst. Civ. Eng. Constr. Mater. 2019; 172 (2): 85-94. https://doi.org/10.1680/jcoma.17.00031.

[9]U. Sivasankaran, S. Raman, S. Nallusamy. Experimental analysis of mechanical properties on concrete with nano silica additive. J. Nano Res. 57 (2019) 93-104.https://doi.org/10.4028/www.scientific.net/JNanoR.57.93.

[10] E. Ghafari, H. Costa, E. Júlio. A. Portugal, L. Durães. The effect of nano-silica addition on ultra-high performance concrete's flowability, strength, and transport properties. Mater. Des. 59 (2014) 1-9. https://doi.org/10.1016/j.matdes.2014.02.051

[11] Jalal M, Pouladkhan A, Harandi OF, Jafari D. Comparative study on effects of Class F fly ash, nano silica, and silica fume on properties of high-performance self-compacting concrete. Constr. Build. Mater. 94 (2015) 90-104. https://doi.org/10.1016/j.conbuildmat.2015.07.001

[12] SANS 50197-1:2013. Cement, Part 1, Composition, specifications and conformity criteria for common cements.

[13] SANS 50196-3:2006. Methods of testing cement part 3. Determination of setting times and soundness.

[14] ASTM C305-13 Standard Practice for Mechanical Mixing of Hydraulic cement pastes and mortars of plastic consistency.

[15] SANS 3001-AG23:2014. Civil engineering test methods. Part AG23, Particle and relative densities of aggregates.

[16] SANS 3001-AG1:2014. Civil engineering test methods. Part AG1, Particle size analysis of aggregates by sieving.

[17] S. Chowdhury, M. Mishra, and O. Suganya. The incorporation of wood waste ash as a partial cement replacement materials for making structural grade concrete: An overview. Ain Shams Eng. J. (2014) https://dx.doi.org/10.1016/j.asej.2014.11.005

[18] L. Skevi, V.A. Baki, Y. Feng, M. Valderrabano, and X. Ke. Biomass bottom ash as supplementary cementitious material: the effect of mechanochemical pre-treatment and mineral carbonation. Materials. 15 (2022) https://doi.org/10.3390/ma15238357.

[19] S. Hailong, Z. Xiuzhi, Z. Peng, and L. Di. Effects of nanosilica particle size on fresh state properties of cement paste. KSCE Journal of Civil Engineering. 25 (2021) 2555 – 2566. Doi.10.1007/s12205-021-0902-3

[20] A. Pratibha, P.S. Rahul, and A. Yogesh. Use of nanosilica in cement based materials. A review. Cogent Engineering, 2 (2015). https://doi.org/10.1080/23311916.2015.1078018

[21] Y. Qing, Z. Zenan, K. Deyu, and C. Rongshen. Influence of nano-SiO2 addition on properties of hardened cement paste as compared with silica fume. Construction and Building Materials. 21 (2007) 539-545.

[20] L. Senff, D. Hotza, W.L. Repette, V.M. Ferreira, and J.A. Labrincha. Mortars with nano-SiO2 and micro-SiO2 investigated by experimental design. Construction and Building Materials. 24 (2010) 1432–1437. http://dx.doi.org/10.1016/j.conbuildmat.2010.01.012

[21] L. Senff, J.A. Labrincha, V.M. Ferreira, D. Hotza, and W.L. Repette. Effect of nano-silica on rheology and fresh properties of cement pastes and mortars. Construction and Building Materials. 23 (2009) 2487–2491. http://dx.doi.org/10.1016/j.conbuildmat.2009.02.005

[22] Kawashima S, Hou P, Corr DJ, Shah SP. Modification of cement-based materials with nanoparticles. Cement and Concrete Composites. 36 (2012) 8-15.

[23] L.P. Singh, S.R. Karade, S.K. Bhattacharyya, M.M. Yousuf, and S. Ahalawat. Beneficial role of nanosilica in cement-based materials – A review. Construction and Building Materials. 47 (2013) 1069-1077. doi.10.1016/j.conbuildmat.2013.05.052

Advances in Cement and Concrete
Materials Research Proceedings 51 (2025) 10-19

Materials Research Forum LLC
https://doi.org/10.21741/9781644903537-2

Characterization of rice husk ash obtained from five rice producing companies in Nigeria

Ayinde Akeem RAHEEM[1,a], Taofiq BELLO[2,b*]

[1]Department of Building, Ladoke Akintola University of Technology, Ogbomoso, 212102, Nigeria

[2]Nigerian Building and Road Research Institute, Km. 10, Idiroko Road, Ota, 112212, Nigeria

[a]aaraheem@lautech.edu.ng, [b]taofiqbello4@gmail.com

Keywords: Rice Husk Ash, Rice Processing Mill, Chemical Composition, Silica, Pozzolan, Nigeria

Abstract. This study analyzed Rice Husk Ash (RHA) from five large rice mills in Nigeria using X-ray fluorescence (XRF) and X-ray diffraction (XRD) to determine their chemical composition and suitability for construction. The collected samples were ground and tested. XRF results showed that all samples contained over 70% silica (SiO_2), specifically at 70.11%, 81.14%, 78.65%, 77.51%, and 78.05% for SP1, SP2, SP3, SP4, and SP5 respectively. Given the content of SiO_2 and Loss on Ignition (LOI), only SP2 aligns with the conditions of class N pozzolan as described in ASTM C 618 (2019). XRD analysis indicated that all samples contained an amorphous phase at different compositions, most of which is beneficial for use as supplementary cementitious material (SCM). According to the XRD results, all samples exhibited the presence of quartz, which corroborates the siliceous nature of these materials as indicated by the XRF results. Despite the variations observed from characterization results, the RHAs demonstrated good pozzolanic properties according to ASTM C 618 (2019). The study concluded that rice processing mills offer RHA as a ready-made pozzolan material that may be applicable in the construction industry for cement production; this can potentially reduce carbon footprints in this sector. This also provides a viable waste management option for RHA disposal while improving cement properties.

Introduction

Nigeria's growing population and increasing income levels have led to a rising demand for rice, a staple product [1]. The country is the largest importer of rice worldwide, the top consumer, and one of its leading producers [2]. Rice consumption in Nigeria stands at 32 kilograms per person annually, reflecting a 4.7% increase over the last ten years according to [3]. The government has prioritized boosting domestic rice production, reaching a peak of 3.7 million tonnes in 2017 [4]. STATISTA (2023) examined annual rice production in Nigeria over a 12-year period from 2010 to 2022, indicating consistent growth as illustrated in Figure 1. However, only 30% of Nigeria's irrigable rice fields are currently under irrigation; thus far, irrigation water provision remains underdeveloped [6]. With over 79 million hectares of cultivated land available for agriculture, only about 39% is utilized for rice production. This situation presents significant potential for expanding rice output if irrigation farming practices are maximized and production becomes fully mechanized with government support. While there is potential for scaling up rice production through improved irrigation farming and mechanization with government assistance [3], this will also lead to increased waste generation from rice processing. These residues pose environmental pollution risks and hold little economic value. Figure 2 depicts a typical RHA dump site from one of the visited mills. The properties of Rice Husk Ash (RHA) vary based on production methods [7]. The source material, Rice Husk (RH) contains approximately 50% cellulose, between 25-30% lignin, and around 14-25% silica [8]. Combustion produces silica ash that can be either crystalline or amorphous; however, crystalline silica is considered less reactive than its amorphous counterpart

Advances in Cement and Concrete
Materials Research Proceedings 51 (2025) 10-19

Materials Research Forum LLC
https://doi.org/10.21741/9781644903537-2

[9]. Controlled burning at temperatures between 500-800°C yields amorphous RHA [7]. Literature suggests that the amorphous silica content in RHA ranges widely: approximately 80-95% [10]; 94-96% [11]; 87-99.77% [12]; 72.6% [13].

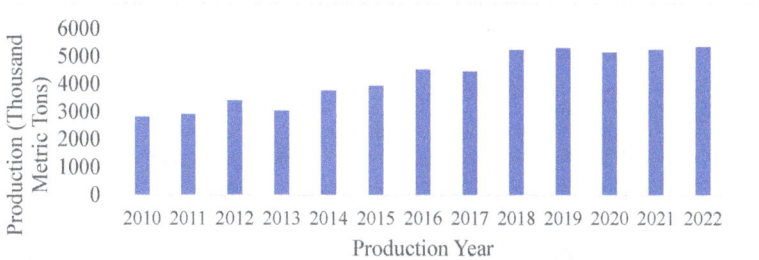

Fig. 1. Annual Rice Production in Nigeria from 2010 to 2022 (1,000 metric tons) Source: [5]

Fig. 2. A Typical RHA Deposit Site in Nigeria

Researchers [13-16], among others, affirmed that RHA possesses high silica content, which is advantageous for cement applications. RHA enhances concrete mix pumpability and reduces cement consumption, thereby lowering both the carbon footprint and costs [15,17]. It holds significant promise for the construction industry in terms of reducing CO_2 emissions. While CO_2 emissions from cement manufacturing contribute to climate change [18], RHA also presents environmental challenges and waste management issues [13]. Singh and Singh (2020) characterized RHA from industrial sources using X-ray fluorescence (XRF) and X-ray diffraction (XRD) techniques, revealing a substantial presence of major oxides as well as silica with amorphous content. Das *et al.* (2022) found that ultra-fine RHA was composed predominantly of silica at 96.2%, accompanied by a prominent quartz peak. Ali *et al.* (2022) assessed the pozzolanic characteristics of two varieties of RHA, reporting silica contents of 78.16% and 73.30%, both exhibiting amorphous structures. The efficacy of XRF and XRD techniques has been demonstrated in characterizing RHA, confirming its potential for use in the construction sector, particularly in cement products. However, there is a notable lack of comprehensive characterization studies on Nigerian RHA derived directly from local rice processing mills. This study aimed to evaluate RHA collected from selected major rice processing mills across Nigeria using X-ray fluorescence and X-ray diffraction analysis methods to examine their elemental composition and mineralogical features with the objective of generalizing their use as a cementitious material for global economic benefits, especially within the construction industry. This approach will help reduce escalating issues related to RHA disposal while conserving finite natural resources; it may also mitigate potential environmental damage posed by improper disposal practices associated with both RHA waste management and emissions from the cement industry.

Materials and Methods

This investigation obtained samples of RHA from five large rice processing mills as detailed in Table 1. All mills exhibited similar procedures for producing ash residues: gathering/preparing Rice Husk (RH), transporting it into a bunker for temporary storage, then feeding it into a boiler furnace via a screw feeder system as illustrated in Figure 3. RHA samples were collected from dump sites at each mill location; specifically at five randomly selected spots on each heap's surface at depths not exceeding 50 mm to ensure sample integrity enough for analysis purposes. Samples were stored in airtight polythene bags to maintain dryness until further processing occurred; milling resulted into fine powder forms suitable for chemical characterization, specifically obtaining particle sizes that passed through a 75 μm sieve from each sample [21].

Table 1. Location, Capacity and Temperature Details Selected Rice Processing Mills

S/N	Rice Processing Mill	Location (state)	Capacity (Tons/Hr)	Furnace Operating Temperature (°C)
1	SP-1	Lagos	15	600-700
2	SP-2	Kwara	10	600-700
3	SP-3	Kebbi	10	700-800
4	SP-4	Kebbi	10	600-800
5	SP-5	Kebbi	10	700-800

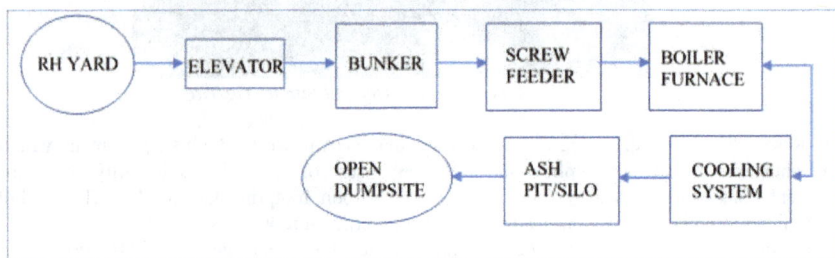

Fig. 3. Typical Operation Processes of RHA Generation in Sampled Rice Companies

Loss on Ignition (LOI) was conducted on five RHA samples following the procedures outlined in ASTM D7348 (2021). The analysis was performed using a muffle furnace capable of reaching a maximum temperature of 1400°C. LOI quantifies the mass loss of a sample when heated to 950°C. For theXRD test, the same samples prepared for XRF analysis were utilized. The examination was carried out using an X-ray diffractometer (XDS 2400H with MiniFlex2 goniometer and detector). The d-spacing for each peak observed in the diffraction patterns was calculated using Bragg's law, and these results were compared against known d-spacings from the International Centre for Diffraction Data (ICDD) database.

Results and Discussions

Analysis of Oxides Compositions

The major oxides that classify a material as a pozzolan (SiO_2, Al_2O_3, and Fe_2O_3) were found in quantities comparable to existing literature records as shown in Table 2. SiO_2 was identified as the dominant component across all RHA samples, with SP2 exhibiting the highest concentration at

84.14%. This was followed by SP3 (78.65%), SP5 (78.05%), SP4 (77.51%), and SP1 (70.11%). The significant variation in SiO_2 content, particularly the elevated level in SP2, may suggest locational or production differences influencing the ash's silica retention capability. SiO_2 is crucial for applications such as concrete production due to its impact on strength and durability [10]. Regarding Al_2O_3 content, SP2 recorded a notably higher concentration at 3.35%, far surpassing that of other samples: SP1 (0.54%), SP3 (0.47%), SP5 (0.27%), and SP4 (0.23%). The elevated alumina levels in SP2 could be attributed to environmental factors like soil conditions [23]. Higher Al_2O_3 content typically enhances pozzolanic reactivity, making RHA more effective in cementitious applications. A similar trend was observed for Fe_2O_3 content; again, SP2 exhibited the highest concentration at 1.02%, compared to lower values from other samples: SP1 (0.49%), SP3 (0.45%), SP5 (0.31%), and SP4 (0.17%). The increased Fe_2O_3 levels in sample SP2 may be linked to geochemical factors [24], which can influence both the ash's color and its magnetic properties.

Other oxides such as calcium oxide (CaO) and magnesium oxide (MgO) showed minimal variation across samples; however, CaO had its highest concentration in sample S(P)2 at 0.43%, closely followed by SP3 at 0.42%. Samples SP1, SP5, and SP4 exhibited lower CaO contents ranging from 0.14% to 0.30%. The presence of CaO enhances the ash's potential as a filler material in construction by improving mechanical properties [25]. In terms of Loss on Ignition (LOI), sample SP1 showed the highest value suggesting greater organic matter or moisture content; conversely, sample SP2 demonstrated thermal stability with an LOI of only 8.22%. All other samples exceeded the recommended LOI threshold of <10% according to [22]. High LOI can adversely affect pozzolanic activity due to reductions in SiO_2 caused by unburnt carbon residues [22]. Elevated LOI levels are often associated with discoloration issues along with problems related to air-entrainment phenomena leading to segregation or reduced compressive strength within mixed components [26]. Such complications typically arise when burning temperatures fall below 600°C [7].

Table 2. XRF Results of the Five RHA Samples in Percentages

Compounds	SP1	SP2	SP3	SP4	SP5
SiO_2	70.11	84.14	78.65	77.51	78.05
Al_2O_3	0.54	3.35	0.47	0.23	0.27
Fe_2O_3	0.49	1.02	0.45	0.17	0.31
CaO	0.3	0.43	0.42	0.14	0.24
MgO	0.16	0	0.13	0.07	0.11
SO_3	0	1.01	0	0.07	0
K_2O	0.52	0	0.4	0.45	0.38
Na_2O	ND	0.37	ND	ND	ND
(LoI)@ 950°c	26.94	8.22	18.76	20.74	20.08

Note: ND means Not Detected

Comparison of RHA Samples with ASTM C 618 (2019) Standard for Class N Pozzolan

The RHA samples showed a combined content of SiO_2, Al_2O_3, and Fe_2O_3 exceeding 70%, as illustrated in Figure 4. The recorded SiO_2 contents were 71.14%, 88.51%, 79.57%, 77.91%, and 78.63% for SP1, SP2, SP3, SP4, and SP5 respectively; notably, SP2 exhibited the highest SiO_2 concentration. Importantly, all samples contained SiO_2 levels greater than the recommended threshold of 70% for Class N pozzolans. Regarding SO_3 content, none of the RHA samples exceeded the maximum limit specified in [22]. The highest SO_3 concentration was found in sample SP2 at 1.01%, followed by SP4 at 0.07%. Samples SP1, SP3, and SP5 did not contain any detectable SO_3. ASTM C618 (2019) recommends an LOI of no more than 10% for Class N pozzolans; however, all samples fell short of this specification except for sample SP2 which demonstrated overall compliance.

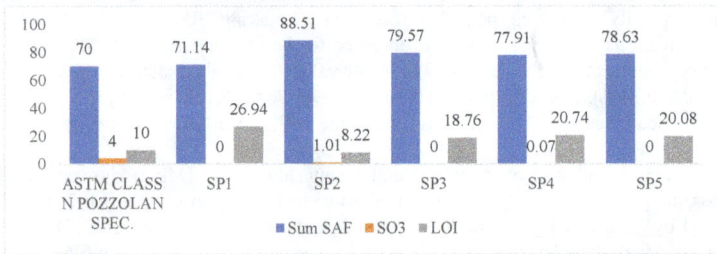

Note: SAF means summation of SiO_2, Al_2O_3 and Fe_2O_3

Fig. 4. Comparison of RHA Samples with ASTM C 618, (2019) for Class N Pozzolan

Statistical Analysis of the Major Oxide Contents of the RHA Samples

To compare the components (SiO_2, Al_2O_3, Fe_2O_3) of SP1, SP2, SP3, SP4, and SP5 against Class N Pozzolan specifications for statistically significant differences at an alpha level of 0.05, Table 3 presents the standard error mean for the five analyzed RHA samples with a P-value of 0.0001. The SiO_2 content exhibited statistically significant differences among the samples (P < 0.05), with SP2 having the highest SiO_2 concentration at 84.14%, which significantly differs from all other samples. Although SiO_2 content in SP3 is lower than that in SP5, this difference is not statistically significant (P > 0.05). Al_2O_3 also showed significant variation between some samples. Here as well, SP2 had the highest alumina content and demonstrated a statistically significant difference compared to others. Conversely, no significant difference was found between SP1 and SP3 or between SP4 and SP5 (P > 0.05). Significant differences were also observed among RHA samples in terms of Fe_2O_3 content; specifically, SP2 recorded the highest value at 1.02%. While both SP1 and SP3 had lower Fe_2O_3 values than that of SP2 , they were not significantly different from each other (P < 0.05). However, sample SP3 (0.45%) was found to have a lower value than sample SP1 (0.49%).

Overall, there was a notable distinction where SP2's Fe_2O_3 level was significantly different from those in SP4, SP5 and combined values for SP1/SP3. The total pozzolanic content (ΣSAF) also exhibited significant differences; while samples SP3, SP4 and SP5 shared relatively similar pozzolanic contents, albeit still significantly distinct due to P < 0.05. SP2 stood out with higher values across all measured parameters: SiO_2, Al_2O_3, Fe_2O_3 ,as well as their summation. Despite exhibiting higher levels of these components, SP2 displayed the lowest LOI (Loss on Ignition) at 8 .22% among all samples examined. SP1 recorded the highest LOI value at 26 .94%, which is significantly different from those of SP3, SP4 and SP5. In contrast, SP4 (20 .74%) and SP5 (20 .08%) showed closely related LOI values but remained statistically distinct as evidenced by P< 0 .05.

Advances in Cement and Concrete
Materials Research Proceedings 51 (2025) 10-19

Materials Research Forum LLC
https://doi.org/10.21741/9781644903537-2

Table 3. Standard Error Mean of the RHA Samples

Specimen	SP1	SP2	SP3	SP4	SP5	P value
SiO_2	70.11±0.33[a]	84.14±0.11[b]	78.65±0.28[c]	77.51±0.23[d]	78.05±0.06[cd]	<0.0001
Al_2O_3	0.54±0.03[a]	3.35±0.02[b]	0.47±0.05[a]	0.23±0.01[c]	0.27±0.02[c]	<0.0001
Fe_2O_3	0.49±0.01[a]	1.02±0.04[b]	0.45±0.01[a]	0.17±0.02[c]	0.31±0.02[d]	<0.0001
$\sum SAF$	71.14±0.32[a]	88.51±0.10[b]	79.57±0.25[c]	77.91±0.23[d]	78.63±0.06[e]	<0.0001
LOI	26.94±0.040[a]	8.22±0.105[b]	8.76±0.061[c]	20.74±0.009[d]	20.08±0.040[e]	<0.0001

Analysis of XRD Results of the RHA Samples

Figure 5a-e illustrates the XRD patterns of five rice husk ash (RHA) samples, denoted as SP1 through SP5. In Figure 5a, the XRD pattern for SP1 reveals a prominent hump in the range of 15°-30° 2θ, indicating significant amorphous material content. This region's lower intensity and absence of sharp peaks are typical for amorphous phases, suggesting that a substantial portion of silica is present in reactive form. However, there are also detectable amounts of crystalline silica. Major peaks at 24.00°, 25.98°, and 28.72° (corresponding to planes 202, 310, and 200 respectively) suggest the presence of crystalline graphite (ICDD, 2004), which may result from incomplete combustion during processing. Additionally, quartz, a stable silica typically formed at higher temperatures is confirmed by crystalline peaks at approximately 33.14° and 43.24° 2θ (planes 220 and 311). The combined crystallinity indicates that SP1 contains around 46.74 wt% crystalline material; this crystallinity limits its pozzolanic reactivity since crystalline silica is considerably less reactive [7,19].

In Figure 5b, Peaks 1 and 2 appearing at 2θ = 24.51° and 34.98°, attributed to amorphous phases with compositions measuring 50.88 wt% and 22.39 wt%, respectively. The broad peak observed at 24.51° 2θ suggests a high degree of amorphosity within SP2 with an overall amorphous/glassy content reaching approximately 73.27 wt %, enhancing its reactivity for pozzolanic applications. This sample also exhibits a peak at 42.98° 2θ corresponding to graphite due to incomplete combustion during organic material processing [26]. The peak observed at 50.26° 2θ relates to hematite (Fe₂O₃) indicating iron oxide presence possibly sourced from impurities or conditions during ash formation. The presence of calcite (CaCO₃) was noted at 57.00° 2θ, likely due to atmospheric carbon dioxide reacting with any calcium oxide (CaO) present in SP2.

Similarly, in Figure 5c, both amorphous and crystalline phases were observed. A broad hump between15°-30° 2θ corresponds roughly to 67.77 wt % of the total composition being amorphous and indicative of glassy silica. Additionally, the most intense peak at around 24.5° 2θ (plane 111) suggests some form of crystalline silica, such as cristobalite or minor quartz; while the lesser-intense peak at around 27° 2θ (plane 310) indicate secondary crystal impurities. Further peaks noted at planes 200 ,220 ,311 and 222 corresponding to 35°, 42.5°, 50.28° and 56.17° 2θ respectively correspond to lower intensity trace components including hematite, graphite and calcite similar to [27].

The XRD pattern presented in Figure 5d shows that SP4 also exhibits a high degree of amorphous content accompanied by a few crystalline phases. The broad hump around 15–30° 2θ indicates significant non-crystalline phase considered highly desirable as a SCM in cement applications. SP4 contains about 37.15 wt% of crystalline silica thus confirming it possesses approximately 62.85 wt% of amophorous silica in its total composition.

Finally, Figure 5e shows the XRD pattern of SP5 and it is also reflecting both amorphous and crystalline characteristics, with a broad hump at lower angles and sharp peaks at higher angles. The broad hump between 15° and 30° 2θ indicates the presence of an amorphous phase. The distinct peaks (111, 202, 310, 200, 220, 311, and 222) overlaid on the amorphous hump and at higher angles signify the presence of crystalline phases corresponding to the presence of graphite,

Advances in Cement and Concrete

Materials Research Forum LLC

Materials Research Proceedings 51 (2025) 10-19

https://doi.org/10.21741/9781644903537-2

hematite, quartz and calcite according to the ICDD, 2004 database. Glassy amorphous silica content represents about 61.16 wt% content indicating that SP5 is rich in amorphous silica as indicated by the broad peaks than crystalline silica indicated by the few minimal peaks, hence reactive and suitable for use as SCM. Figure 5e depicts a pattern of prominent amorphous hump and multiple crystalline peaks, suggesting a high pozzolanic potential due to the amorphous silica content.

Summarily, there were distinct differences in the crystallographic structure of the sampled RHAs, that is, SP1-5 corresponding to Figure 5a-e. Generally, all the RHAs contains significant amorphous silica desirable as a supplementary cementitious material especially in the cement industry. However, SP2 contains the highest quantity of amorphous silica at 73.27 wt% compared to 53.26 wt%, 67.77 wt%, 62.85 wt% and 61.16 wt% of SP1 (Figures 5a), SP3 (Figure 5c), SP4 (Figure 5d) and SP5 (Figure 5e) respectively. The presence of significant content of reactive silica in these samples will drive reactivity [7, 19], and the less crystalline phases will influence the ash's physical characteristics such as colouration.

Generally, RHA from rice processing industries appears to be desirable for supplementary cementitious applications. The differences in crystallographic structures of these RHA samples may suggests difference in operating temperatures and efficiency of the boiler furnace, handling or exposure conditions of the RH before combustion and RHA after combustion when discharged to the open field. The presence of impurities such as hematite and calcite was found typical across board, and may be curtailed if the rice processing industry carefully handled and stored the materials (RH and RHA). Similarly, the presence of crystalline silica in form of quartz was evidenced in trace quantity across board indicating that the samples experienced a high treatment temperature of likely more than 750°C or 800°C for prolong period [7] and this can be possible after discharging RHA from the furnace to where there is no further temperature regulation.

Ultimately, all the samples have an aggregation of the major oxides in excess of 70% regardless of their location and this suggest that rice processing mills are potential source of silica-rich RHA suitable for various industrial applications in cement industries. However, SP2 exhibited the most desired pozzolan properties

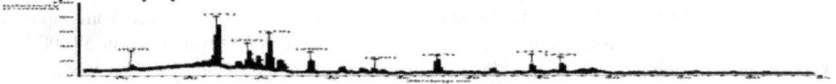

Figure 7a. XRD Pattern of SP1

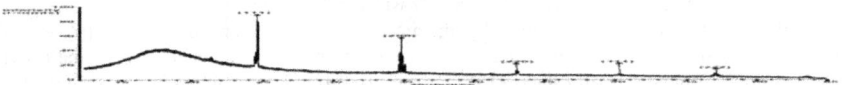

Figure 7b. XRD Pattern of SP2

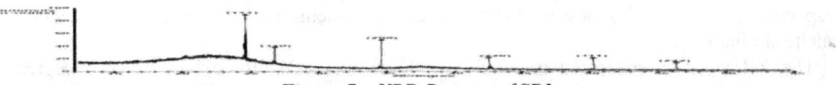

Figure 7c. XRD Pattern of SP3

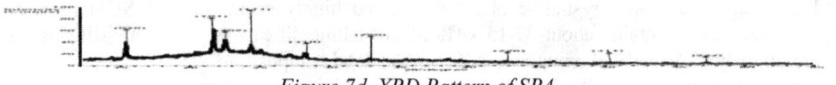

Figure 7d. XRD Pattern of SP4

Figure 7e. XRD Pattern of SP5

Advances in Cement and Concrete
Materials Research Proceedings 51 (2025) 10-19

Materials Research Forum LLC
https://doi.org/10.21741/9781644903537-2

Comparison with Past Studies

The SiO_2 content in RHA is a crucial factor when selecting it as a SCM especially in the amorphous phase is desirable for high-reactivity applications, such as in cement. The XRF analysis of the five RHA samples from this study generally shows amorphous phase of SiO_2 levels comparable to both local and international findings. For example, the SiO_2 content in SP1 (70.11%) closely aligns with the results reported by [29,13, 20]. Similarly, SP2, with 84.14% silica, mirrors findings from [16], while SP3 and SP5, with 78.65%, 77.51%, and 78.05% SiO_2, are comparable to [20].

The XRD analysis of the RHAs revealed an amorphous phase, which when combined with the obtained SiO_2, Al_2O_3 and Fe_2O_3 from the XRF analysis suggests their strong potential as SCMs in the construction industry. The high silica content of the RHA samples analyzed in this study showed that the material holds a great promise as a SCM in cement production [30]. Utilizing RHA as a SCM in cement production can contribute to the eco-friendliness in the cement industry [16]. RHA can pose environmental and health risks, so finding a sustainable use for this residue is essential. This study analyzed RHA from five major rice processing mills in Nigeria, located in Lagos, Kwara, and Kebbi states, to assess their potential for use in the construction industry.

Conclusions

i. RHA from different rice processing mill sources showed SiO_2, Al_2O_3 and Fe_2O_3 in quantities that is suitable for use as supplementary cementitious material in the construction industry,

ii. There were significant variations in the chemical compositions and crystallographic structure of the RHA samples attributable to differences in production temperature and exposure condition,

iii. RHA from rice processing industries exhibited amorphous structure, hence reactive and suitable for cement application and offering a good waste management option for RHA disposal while improving the properties of cement.

References

[1] B. P. Ekundayo, Rice Production, Imports and Economic Growth In Nigeria: An Application of Autoregressive Distributed Lag. International Journal of Advanced Economics 5 (2) (2023) 48-56. https://doi.org/10.51594/ijae.v5i2.449

[2] G. Okpiaifo, A. Durand-Morat, G. H. West, L. L. Nalley, R. M Nayga Jr, and E. J. Wailes, Consumers' preferences for sustainable rice practices in Nigeria. Global Food Security 24 (2020) 100345.

[3] Price Waterhouse Coopers (PWC), Boosting rice production through increased mechanization. Accessed: August 2024 from *https://www.pwc.com/ng/en/publications/boosting-rice-production-through-increased-mechanisation.html*

[4] Rice Farmers Association of Nigeria Report on Rice Exportation, RIFAN, Accessed, 29 October, 2021 from *https://www.vanguardngr.com/2021/06/nigeria-poised-to-export-rice-as-local-production-shoots-to-9m-metric-tonnes/*

[5] STATISTA, Production of milled rice in Nigeria from 2010 to 2023. Accessed, 25 August, 2024 from *https://www.statista.com/statistics/1134510/production-of-milled-rice-in-nigeria/*

[6] N. Danbaba, J. C. Anounye, A. S. Gana, M. E. Abo, M. N. Ukwungwu, and M. H. Badau, Grainphysico-chemical and milling qualities of rice (Oryza sativa) cultivated in southwest Nigeria Journal of Applied Agricultural Research, 5 (1) (2013) 61-71.

[7] C. Fapohunda, B. Akinbile, and A. Shittu, Structure and properties of mortar and concrete with rice husk ash as partial replacement of ordinary Portland cement- A review. International Journal of Sustainable Built Environment, 6 (2) (2017) 676-689. https://doi.org/10.1016/j.ijsbe.2017.07.004

[8] A. Bazargan, Z. Wang, J. P. Barford, J. Saleem, and G. McKay, Optimization of the removal of lignin and silica from rice husks with alkaline peroxide. Journal of Cleaner Production 260 . (2020) 120848. https://doi.org/10.1016/j.jclepro.2020.120848

[9] C. L. Hwang, and S. Chandra, The Use of Rice Husk Ash in Concrete. *https://3cl1105uvd.files.wordpress.com/2013/06/book-by-satish-chandra-220504.pdf.* Assessed on the 27th of April, 2016.

[10] S. K. Das, J. Mishra, S. M. Mustakim, A. Adesina, C. R. Kaze, and D. Das, Sustainable utilization of ultrafine rice husk ash in alkali activated concrete: Characterization and performance evaluation. Journal of Sustainable Cement-Based Materials, 11(2) (2022) 100-112. https://doi.org/10.1016/j.conbuildmat.2022.128341

[11] D. Pranowo, A. A. Nuryono, M. R. Jumina, and F. M. C. S. Setyabudi, Application of silica extracted from rice husk ash for the encapsulation of AFB1 antibody as a matrix in immunoaffinity columns. JSM Mycotoxins, 67 (2) (2017) 77-83. https://doi.org/10.2520/myco.67_2_1

[12] A. Putranto, S. Abida, A. Sholeh, and H. Azfa, The potential of rice husk ash for silica synthesis as a semiconductor material for monocrystalline solar cell: A Review. IOP Conf. Ser.: Earth Environ. Sci. 733 (1) (2021) 012-029. https://doi.org/10.1088/1755-1315/733/1/012029

[13] A. A. Raheem, and M. A. Anifowose, Effect of ash fineness and content on consistency and setting time of RHA-blended-cement. Materials Today: Proceedings 86 (2023) 18-23. https://doi.org/10.1016/j.matpr.2023.02.054

[14] H. B. Mahmud, S. Bahri, Y. W. Yee, Y. Y. Yeap, Effect of rice husk ash on the strength and durability of high strength performance concrete. World Acad. Sci. Eng. Technol. 10 (3) (2016) 375–380.

[15] J. Kamau, A. Ahmed, and K. Ngong, Sulfate Resistance of Rice Husk Ash Concrete. MATEC Web of Conferences 199 (2018) 02006. https://doi.org/10.1051/matecconf/201819902006

[16] A. Z. Seyed, A. Farshad, D. Farzan and A. Mojtaba, Rice husk ash as a partial replacement of cement in high strength concrete containing micro silica: Evaluating durability and mechanical properties. Case Studies in Construction Materials 7 (2017) 73-81, https://doi.org/10.1016/j.cscm.2017.05.001.

[17] V. Venkitasamy, M. Santhanam and B. P. C. Rao, Workability Assessment of Pumpable Structural-Grade Heavy-Weight Concrete Using Novel Coaxial Cylinder Method. Journal of Materials in Civil Engineering, 35 (7) (2023) 04023197. https://doi.org/10.1061/JMCEE7.MTENG-14311

[18] J. Skibsted and R. Snellings, Reactivity of supplementary cementitious materials (SCMs) in cement blends. Cement and Concrete Research, 124 (2019) 105799.

[19] A. Singh and B. Singh, Characterization of rice husk ash as obtained from industrial source. *Journal of Sustainable Cement-Based Materials,* 10 (4) (2020) 193-212. https://doi.org/10.1080/21650373.2020.1789010

Advances in Cement and Concrete
Materials Research Proceedings 51 (2025) 10-19

Materials Research Forum LLC
https://doi.org/10.21741/9781644903537-2

[20] D. Ali, U. A, Jalam and M. Balteh, Pozzolanic Characterization of Rice-Husk-Ash from Two Varieties of Rice in Gombe: A Strategy for Achieving Low-Cost Housing. International Journal of Innovative Science and Research Technology 7 (11) (2022) 767-772.

[21] American Society for Testing Materials, ASTM E1621 (2021). Standard Guide for Elemental Analysis by Wavelength Dispersive X-Ray Fluorescence Spectrometry. ASTM International, 100 Barr Harbor Drive, PO Box C700, West Conshohocken, PA, 19428-2959 USA

[22] American Society for Testing Materials, ASTM C618 (2019). Standard Specification for Coal Fly Ash and Raw or calcined Natural pozzolan for use in concrete. *ASTM Inc.* in US

[23] B. H. J. Pushpakumara and W. S. W. Mendis, Suitability of RHA with lime as a soil stabilizer in geotechnical applications. International Journal of Geo-Engineering 13 (1) (2022). 4.

[24] J. Grabic, M. Vraneaievia, R. Zemunac, S. Zdero, A. Bezdan and M. Ilic, Iron and Manganese in well water: Potential risks for irrigation system. Acta Horticultureet Regiotecturae 22 (2019) 93-96.

[25] M. H. El-Ouni, A. Raza, H. Haider, M. Arshad and B. Ali, Enhancement of mechanical and toughness properties of carbon-fibre reinforced geopolymer pastes comprising calcium nano-calcium oxide. Journal of the Australian Ceramic Society 58 (4) (2022) 1375-1387.

[26] N. Chuslip, C. Jaturapi and K. Kiattikomol, Effect of LOI of ground bagasse ash on the compressive strength and sulfate resistance of mortars. Construction and Building Material 23 (12) (2009) 3523-3531.

[27] D. J. Rodriguez, D, C. Y. Lau, B. A. Long, S. A. Tang, A.M. Friesa and S. L. Anderson, O2-Oxidation of individual graphite and graphene nanoparticles in 1200-2200 K range: Particle-to-particle variations and evolution of the reaction rates and optical properties, Carbon 173 (2021) 286-300. https://doi.org/10.1016/j.carbon.2020.10.053.

[28] S. H Kang, S. G. Hong and J. Moon, The use of rice husk ash as reactive filler in ultra-high-performance concrete. Cement and Concrete Research 115 (2019) 389-400. https://doi.org/10.1016/j.conbuildmat.2019.04.091

[29] D. G. K. Adamu, U. M. Dankawu, M. N. Maharaz, E. N. Chifu, S. S. Zarma, N. W. Silkwa, and M. Ahmadu, Evaluation of chemical compositions of rice husk from local rice species using x-ray fluorescence technique. Fudma Journal of Sciences 7 (4) (2023) 82-89. https://doi.org/10.33003/fjs-2023-0704-1777

[30] J. Lin, E. Shamsaei, F. B. De Souza, K. Sagoe-Crentsil, and W. H. Duan, Dispersion of graphene oxide–silica nanohybrids in alkaline environment for improving ordinary Portland cement composites. Cement and Concrete Composites 106 (2020) 103488. https://doi.org/10.1016/j.cemconcomp.2019.103488.

Advances in Cement and Concrete
Materials Research Proceedings 51 (2025) 20-28

Materials Research Forum LLC
https://doi.org/10.21741/9781644903537-3

Evaluation of compressive strength of bio-fibrous concrete with silica fume

Muyideen ABDULKAREEM[1,a], Fadilat AYERONFE[2,b],
Abideen GANIYU[3,c] and Wasiu AJAGBE[4,d]

[1]Civil Engineering and Quantity Surveying Department, Military Technological College, Oman

[2]Department of Microbiology, University of Ilorin, Ilorin, Nigeria

[3]Department of Civil Engineering, British University of Bahrain, Bahrain

[4]Department of Civil Engineering, University of Ibadan, Ibadan, Nigeria

[*a]muyikareem@gmail.com, [b]oluwatosinfadu@gmail.com, [c]a.ganiyu@bub.bh,
[d]ajagbewas@gmail.com

Keywords: Bacillus Subtilis, Bio-Fibrous Concrete, Compressive Strength, Kenaf Fiber, Silica Fume

Abstract. Concrete is a widely used material in construction due to its advantageous properties. However, the formation of cracks in concrete presents a significant challenge to its durability and can lead to a decrease in the service life of structures. Researchers have proposed various solutions to address this issue, including the incorporation of bacteria for self-healing and the use of kenaf fibre, and silica fume in concrete. In this study, bacillus subtilis bacteria is utilized to investigate their compressive strength on concrete, while kenaf Fiber and silica fume were added to concrete mixes at different concentrations of 0%, 0.1%, 0.2%, 0.5% and 0%, 1%, and 1%, 2%, 5% and 10%, respectively, to assess its effects on concrete strength. Concrete samples were cast in cube moulds with water-cement ratios of 0.5 and compressive strength testing was conducted on the 28th day to determine the samples' impact resistance. The inclusion of bacillus subtilis in the concrete mix increased the compressive strength of concrete by 6.7%. Furthermore, incorporation of kenaf Fiber and silica fume increases the compressive strength of bio-concrete by about 55%.

Introduction

Over the past few centuries, concrete has been extensively applied as the main construction materials. This is due to its appealing strength and other engineering properties. The performance of these concrete structures is often affected due to overloading as well as the harsh environmental conditions they are exposed to [1, 2]. Based on this, concrete-modifying and property-improving materials are being added to concrete to improve mechanical properties.

Addition of mineral admixtures such as fly ash, rice husk ash, palm oil fuel ash and silica fumes have been shown to improve concrete properties [3, 4]. The drying shrinkage and porosity of concrete decrease with fly ash increment, thus reducing chloride permeability and water sorptivity [5]. Incorporating fly ash in concrete increases the compressive strength by 15%, modulus of elasticity by 15% and plastic deformations by 38% [6]. Rice husk ash increases concrete strength and durability, as well as decreases the corrosion effect caused by harsh exposure conditions [7, 8]. Similar trends have been shown with the addition of rice husk ash to increase the tensile strength and flexural strength as observed by Al-Alwan et al. [9]. Similarly, Okashah et al. [10] showed in their experiment that the compressive strength of concrete increased by 37% by using the optimum portion of silica fume.

Aside using the above mineral admixtures, the past few years have seen researchers applying microorganisms to improve concrete [11, 12]. This is commonly done by using calcite-producing bacteria via metabolism. This bacteria act as a self-healing agent as well as a strength-enhancing

Advances in Cement and Concrete
Materials Research Proceedings 51 (2025) 20-28

Materials Research Forum LLC
https://doi.org/10.21741/9781644903537-3

agent via filing-up of voids in the concrete matrix. For example, Abdulkareem et al. [13] showed that bacillus subtilis bacteria increased the compressive strength of concrete by 130% provided the curing conditions were favourable (humidity, temperature, wind speed and sunlight exposure). Similar increment was reported by Kishore et al. [14] as they increased concrete strength by 20% via using microbes. Sultan et al. [15] presented a study that showed bio-concrete ability to seal up crack width of 0.8 mm, while increasing the compressive strength by 35%.

In this study, the combined benefits of calcite-producing bacteria, kenaf fibre and silica fume are harnessed to improve concrete properties. Calcite-producing bacillus subtilis bacteria is at a concentration of 109 cells/ml together with different contents of both kenaf fibre and silica fume are added to concrete. Kenaf fibre at 0.1%, 0.2% 0.5% and 1% contents are utilised while silica fume are at 1%, 2%, 5% and 10% replacement of the binder (cement). In addition, two control samples (a plain concrete sample and a bio-concrete sample) with neither kenaf fibre nor silica fume are also prepared as a benchmark for this study. Concrete cubes of 100 mm are prepared with 0.5 water-cement (w/c) ratio and cured. These cubes are thereafter crushed at the 28th day to measure the compressive strength.

Materials and Testing
Cement and aggregates. The materials used in this experiment are purchased in Kuala Lumpur. The cement used is Ordinary Portland cement produced by Cement Industries of Malaysia (CIMA) that conforms to BS12 1996 specifications. Fine aggregate applied is river sand with density of 1750 kg/m^3 that passed through sieve 4.75 mm diameter. The coarse aggregate is crushed igneous rock and has a maximum size of 14 mm. For this experiment, a mix ratio of 1:2:2.5 in terms of cement, fine aggregate and coarse aggregate, respectively, is applied.

Bacillus subtilis bacteria. Bacillus subtilis bacteria is prepared by using culture media that contains 5.0 g yeast extract, 10 g tryptone, and 10.0 g NaCl in 950 ml H$_2$O. Fig. 1 shows a cultured bacteria applied in this study.

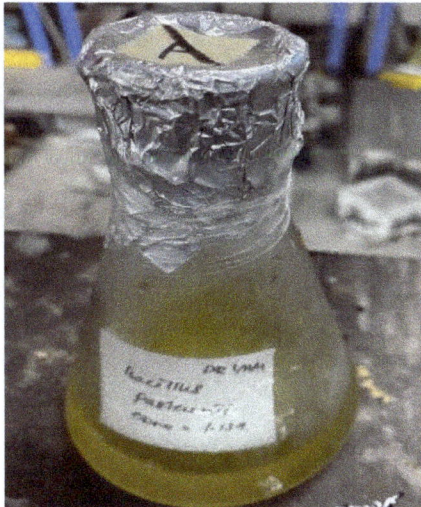

Figure 1: Bacillus subtilis bacteria

Kenaf fibre. The kenaf fibre used is locally purchased fibre made by the National Kenaf and Tobacco Board (LKTN), Malaysia. The fibre was treated before usage with sodium hydroxide (NaOH) to reduce its degradation rate and water sorptivity property [16]. The fibre strands are separated by using a metal comb and are cut to a length of 50 cm. The content of kenaf fibre utilised in this study are 0.1%, 0.2% 0.5% and 1%. Table 1 shows the physical properties of the kenaf fibre. Fig. 2 shows the cut and treated kenaf fibre.

Silica fume. The silica fume (SF) applied is commercially produced by Elkem Carbon Malaysia Sdn. Bhd. The silica fume is in a dry densified state with Grade 920 and conforms to the requirements of BS EN 13263-1 [17]. SF has a surface area of 17950 m^2/kg, and its spherical shape particles significantly aid flowability of concrete during production. Silica fume is added as a partial replacement at 1%, 2%, 5% and 10% by mass of cement. Fig. 3 shows a sample of the silica fume applied in this study. The chemical and physical compositions of the SF are given in Table 2.

Table 1: Physical properties of kenaf fibre

Physical properties		Mechanical properties	
Length (mm)	50	Tensile strength (N/mm²)	930
Density (g/cm³)	1.5	Elastic modulus (GPa)	53
Diameter (μm)	65	Elongation at yield (%)	1.6

Figure 2: Treated kenaf fibre

Table 2: Chemical constituents and physical properties of silica fume fibre

	Chemical constituents								Physical characteristics
Composition	SiO₂	Al₂O₃	Fe₂O₅	CaO	MgO	Na₂O	K₂O	SO₃	Specific gravity
Content (%)	92.2	0.82	0.67	0.64	1.24	-	4.02	0.41	2.2

Advances in Cement and Concrete
Materials Research Proceedings 51 (2025) 20-28

Materials Research Forum LLC
https://doi.org/10.21741/9781644903537-3

Figure 3: Silica fume

Test specimens. A mix ratio of 1:2:2.5 with a water-cement ratio of 0.5 is applied in this study, this is based on results obtained in previous studies. A total of 36 concrete cubes of 100 mm dimensions are cast cured and tested to obtain compressive strength. The bacteria is incorporated into the concrete mix by pouring the bacteria culture into the mixing water. The coarse aggregate is poured into the mixing machine, thereafter, the kenaf fibre is added by separating the fibres. This is done in bits to prevent the kenaf fibre from agglomerating. The fine aggregate, cement and mixing water are thereafter poured into the mixing machine. Two concrete samples are cast for each mix, and the average of their readings is taken to reduce error. Fig. 4 shows some samples of the cast concrete cubes. The samples are demoulded after 24 hrs and cured in a tank. The samples are taken from the tank after 28 days and crushed by using a universal testing machine (UTM) to measure the compressive strength. The UTM used in this study is shown in Fig. 5.

Figure 4: Concrete cube specimens

Figure 5: Compressive strength testing of specimens

Compressive strength

The compressive strength findings obtained on the 28th day of curing are depicted in the Figs. 6-10. Fig. 6 shows the compressive strength of the two control samples (concrete without bacteria and concrete with bacteria). The compressive strength increased from 18 N/mm^2 to 19.2 N/mm^2 when bacteria is added to the concrete sample.

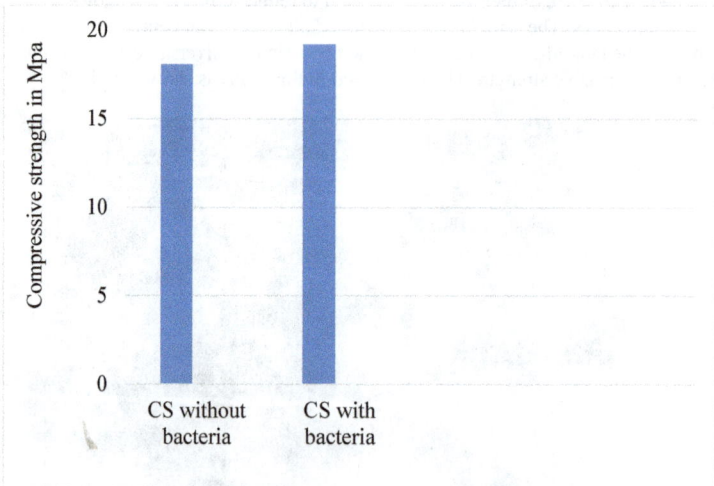

Figure 6: Compressive strength of control samples (CS)

Advances in Cement and Concrete
Materials Research Proceedings 51 (2025) 20-28

Materials Research Forum LLC
https://doi.org/10.21741/9781644903537-3

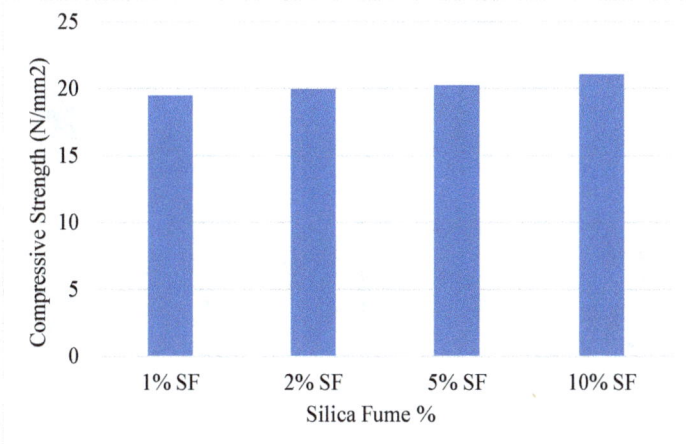

Figure 7: Compressive strength of concrete samples with Bacillus subtilis and 0.1% fiber content

From Fig. 7-10, the addition of silica fume and kenaf fibre increased the compressive strength of concrete. For example, in Fig. 7, the compressive strength of concrete increased from 19.5 N/mm^2 to 21 N/mm^2 when silica fume is increased from 1% to 10% at 0.1% kenaf fibre. Similarly, in Fig. 10, the compressive strength of concrete increased from 27.5 N/mm^2 to 29.8 N/mm^2 when silica fume is increased from 1% to 10% at 1% kenaf fibre.

In addition to the above, Figs. 7-10 show that the compressive strength of concrete increases as the kenaf fibre content increases. The compressive strength of concrete increased from 19.5 N/mm^2 (Fig. 7) to 27.5 N/mm^2 (Fig. 10) when kenaf fibre is increased from 0.1% to 1% at 1% silica fume. Similarly, the compressive strength of concrete increased from 21 N/mm^2 (Fig. 7) to 29.7 N/mm^2 (Fig. 10) when kenaf fibre is increased from 0.1% to 1% at 10% silica fume.

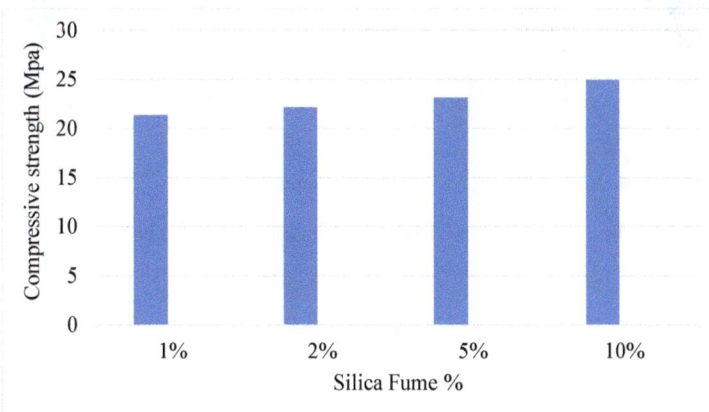

Figure 8: Compressive strength of concrete samples with Bacillus subtilis and 0.2% fiber content

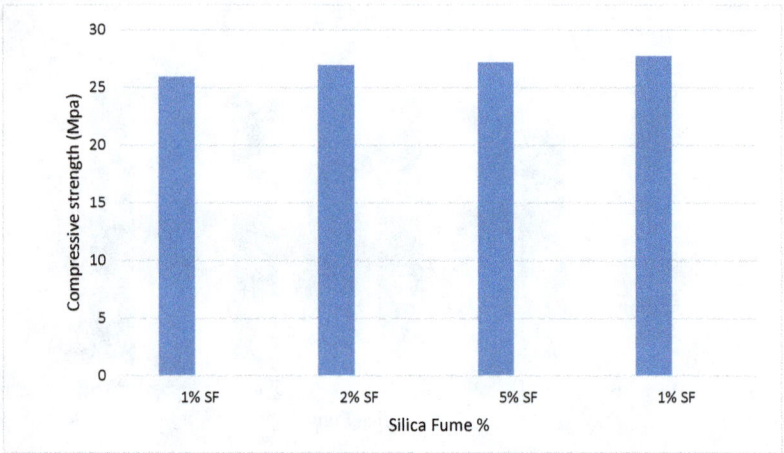

Figure 9: Compressive strength of concrete samples with Bacillus subtilis and 0.5% fiber content

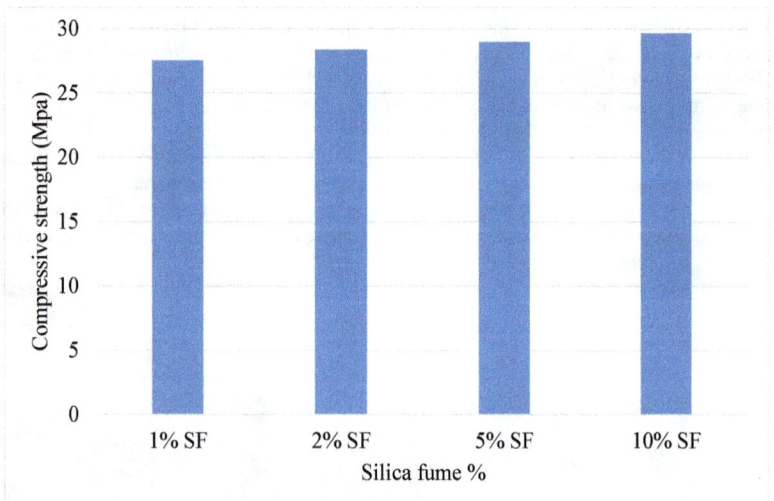

Figure 10: Compressive strength of concrete samples with Bacillus subtilis and 1% fiber content

Advances in Cement and Concrete
Materials Research Proceedings 51 (2025) 20-28

Materials Research Forum LLC
https://doi.org/10.21741/9781644903537-3

Conclusions

The study evaluates the compressive strength of bio-fibrous concrete incorporated with silica fume. The bacteria used in this investigation was bacillus subtilis, and the fibre is Kenaf. Kenaf fibre at 0.1%, 0.2%, 0.5% and 1%, silica fume at 1%, 2%, 5% and 10% are applied concrete at 0.5 w/c ratio. The results showed that incorporation of silica fume and/or kenaf fibre to bioconcrete increases the compressive strength. The results also showed that incorporation of kenaf fibre has more impact in increasing the compressive strength of bioconcrete than silica fume.

References

[1] H.M. Allujami, M. Abdulkareem, T.M Jassam, R.A. Al-Mansob, J.L. Ng, A. Ibrahim, Nanomaterials in recycled aggregates concrete applications: mechanical properties and durability. A review, Cogent Engineering, 9 (2022), https://doi.org/10.1080/23311916.2022.2122885

[2] P.L. Shafigh, J. Chai, H.B. Mahmud, M.A. Nomeli, A comparison study of the fresh and hardened properties of normal weight and lightweight aggregate concretes, Journal of building Engineering, 15 (2018). https://doi.org/10.1016/j.jobe.2017.11.025

[3] F. Aslam, O. Zaid, F. Althoey, S.H. Alyami, S.M. Qaidi, J. de Prado Gil, R. Martínez-García R, Evaluating the influence of fly ash and waste glass on the characteristics of coconut fibers reinforced concrete, Structural Concrete, (2022). https://doi.org/10.1002/suco.202200183

[4] K. Onyelowe, G. Alaneme, C. Onyia, D. Bui Van D, E. Nnadi, C. Ogbonna, L. Odum, D. Aju, C. Abel, I. Udousoro, Comparative modeling of strength properties of hydrated-lime activated rice-husk-ash (HARHA) modified soft soil for pavement construction purposes by artificial neural network (ANN) and fuzzy logic (FL), Jurnal Kejuruteraan, 33 (2021). https://doi.org/10.17576/jkukm-2021-33(2)-20

[5] A.K. Saha, Effect of class F fly ash on the durability properties of concrete. Sustainable environment research, 28 (2018). https://doi.org/10.1016/j.serj.2017.09.001

[6] Z. Rudžionis, E. Ivanauskas, Investigations into effective fly ash used in concrete, Journal of Civil Engineering and Management, 10 (2004). https://doi.org/10.1080/13923730.2004.9636323

[7] M.N. Amin, W. Ahmad, K. Khan, M.M. Sayed, Mapping research knowledge on rice husk ash application in concrete: A scientometric review. Materials, 15 (2022). https://doi.org/10.3390/ma15103431

[8] X. Hong, J.C. Lee, J.L. Ng, M. Abdulkareem, Z.M. Yusof, Q. Li, Q. He, Prediction Model and Mechanism for Drying Shrinkage of High-Strength Lightweight Concrete with Graphene Oxide. Nanomaterials, 13 (2023). https://doi.org/10.3390/nano13081405

[9] A.A.K. Al-Alwan. M. Al-Bazoon, F.I. Mussa, H.A. Alalwan, M.H. Shadhar, M.M. Mohammed, M.F. Mohammed, The impact of using rice husk ash as a replacement material in concrete: An experimental study. Journal of King Saud University-Engineering Sciences, 36 (2022). https://doi.org/10.1016/j.jksues.2022.03.002

[10] A.M. Okashah, M. Abdulkareem, A.Z. Ali, F. Ayeronfe, M.Z. Majid, Application of Automobile Used Engine Oils and Silica Fume to Improve Concrete Properties for Eco-Friendly Construction. Environmental and Climate Technologies, 24 (2020). https://doi.org/10.2478/rtuect-2020-0008

[11] D. Syamsunur, L. Wei, M.N. Hisyam, Z.A. Memon, B. Sultan, Research on Performance Monitoring of Binary Nano Modified Concrete Based on Temperature Variation. Case Studies in Construction Materials, (2023). https://doi.org/10.1016/j.cscm.2023.e02373

[12] S.A. Rizwan, M. Irfan-ul-Hassan, A. Rahim, S. Ali, A. Sultan, D. Syamsunur, N.I. Md Yusoff, Recycled Coarse Aggregate for Sustainable Self-Compacting Concrete and Mortar. Advances in Materials Science and Engineering, (2022). https://doi.org/10.1155/2022/4566531

[13] M. Abdulkareem, F. Ayeronfe, M.Z. Abd Majid, A.R.M. Sam, J.H.J. Kim, Evaluation of effects of multi-varied atmospheric curing conditions on compressive strength of bacterial (bacillus subtilis) cement mortar. Construction and Building Materials, 218 (2019). https://doi.org/10.1016/j.conbuildmat.2019.05.119

[14] K. Kishore, N. Gupta, K.K. Saxena, J. Lade, Development and characterisation of bacteria as a potential application in enduring the mechanical and durability characteristic of cement composite. Advances in Materials and Processing Technologies, 8 (2022). https://doi.org/10.1080/2374068X.2021.1959115

[15] A. Sultan, N. Shaheen, R.A. Khushnood, Experimental evaluation of bacterial self-healing concrete embodying Bacillus pumilus cured in normal and accelerated modes. Materials and Structures, 56 (2023). https://doi.org/10.1617/s11527-023-02129-5

[16] A. Baarimah, S.S. Mohsin, W. Alaloul, M. Ba-Naimoon, Effect of Sodium Hydroxide on Mechanical Characteristics of Kenaf Fibers Reinforced Concrete, Journal of Physics: Conference Series. (2021). doi:10.1088/1742-6596/1962/1/012013

[17] EN, B., 13263-1. Silica fume for concrete–Part 1: Definitions, requirements and conformity criteria. British Standards Institute, London, (2005)

Advances in Cement and Concrete
Materials Research Proceedings 51 (2025) 29-38

Materials Research Forum LLC
https://doi.org/10.21741/9781644903537-4

Characterization and workability of rice husk ash-calcium nitrate blended cement concrete

Akeem Ayinde RAHEEM[1,a], Isaiah Oluwafemi OGUNTOLA[2,b*],
Favour Wuraola KOLAWOLE[2,c] and Oludayo Ajani AKINTOLA[3,d]

[1]Department of Building, Ladoke Akintola University of Technology, Ogbomoso, Nigeria

[2]Nigerian Building and Road Research Institute, Km10, Idiroko road, Ota, Ogun State, Nigeria

[3]Department of Civil Engineering, Ladoke Akintola University of Technology, Ogbomoso, Nigeria

[a]aaraheem@lautech.edu.ng [b]oguntolaisaiah@nbrri.gov.ng
[c]kolawoleoludara@gmail.com [d]oludayo.akintola@gmail

Keywords: Rice Hush Ash, Calcium Nitrate, Supplementary Cementitious Materials, Blended Cement, Workability

Abstract. Blended cement is a type of cement that is produced by blending different supplementary cementitious materials (SCMs) with Portland cement clinker. Ash from agricultural residues which are typically classified as waste, is found to exhibit pozzolanic qualities and are being employed in place of cement in the search for an alternative binder. This study investigates the characterization and workability of Rice Husk Ash (RHA)-calcium nitrate (CN) blended cement concrete. Ordinary Portland cement (OPC), RHA and CN were characterized using X-ray fluorescence analysis. Concrete mixes using a fixed content of 15% RHA as replacement for cement and 1, 2, 3, 4, and 5% CN as substitute by weight of OPC with mix ratio 1:2:4 was produced. The water-to-cement ratio of 0.6 was adopted. Concrete without RHA and CN serves as control 1, while concrete with only RHA serves as control 2 mix. Workability (slump and compacting) and setting times of the concrete were determined. RHA was found to be a pozzolanic material because the sum of SiO_2, AlO_3, and Fe_2O_3 was 75.36% ≥ 70%. The CaO content present in CN, OPC and RHA were 81.95, 68.76 and 5.62%, respectively. The slump and compacting factor ranged from 14 to 34 mm and 0.72 to 0.84 mm, respectively. The initial and final setting time ranged from 129 to 231 minutes and 221 to 379 minutes, respectively. The addition of CN improved the workability and setting times of fresh concrete made with RHA-CN.

Introduction

Rice husk is a by-product generated during the milling of rice. As the outermost layer of the rice grain, it is separated from the grain during the de-husking process. The husk constitutes about 20% of the total weight of the rice [11]. In a country like Nigeria, where rice production is substantial, large quantities of husk are generated annually. The availability of rice husk in Nigeria has also been highlighted as a potential source of pozzolan material [4]. Rice husk can be further processed through controlled burning to produce rice husk ash (RHA), which, when finely ground, acts as a pozzolan in concrete, improving properties such as strength and durability. The controlled combustion process removes organic matter and leaves behind a silica-rich material that reacts with calcium hydroxide in the presence of water to form cementitious compounds [19].

The suitability of RHA for the production of structural concrete, having been classified as highly reactive pozzolan, is not only considered as an environmentally friendly construction material, but also to reduce the cement constituents of concrete production [13]. Rice husk ash (RHA) contains significant amounts of silicon dioxide (SiO_2), making it a SCM that could partially replace cement in concrete [15]. The use of RHA as partial cement replacement in concrete is a

Advances in Cement and Concrete   Materials Research Forum LLC
Materials Research Proceedings 51 (2025) 29-38   https://doi.org/10.21741/9781644903537-4

potential means of developing sustainable binder for construction and is a way of effectively converting the waste to wealth [28].

Studies have demonstrated that calcium nitrate improves the workability of concrete mixes while also preventing segregation and bleeding in concrete that includes pozzolanic materials like rice husk ash, fly ash, and other supplementary materials. Its ability to control the setting time, even under different environmental conditions, ensures the mix maintains sufficient plasticity for proper placement and compaction [31]. Calcium nitrate functions as a setting accelerator and significantly boosts the hydration process in cementitious materials, especially in cement blended with pozzolans. This improved hydration leads to better dispersion of particles and reduced water demand, which, in turn, enhances workability without compromising the strength or durability of the concrete mix [30]. Studies have demonstrated that the addition of CN can help reduce the water demand in RHA concrete mixtures, thereby improving slump values and overall workability [26].

Materials and Methods
The materials used in carrying out this research and the adopted method were discussed in this section.

Materials
Rice Husk Ash (RHA) was sourced from Hillcrest Rice Mill in Offa, Kwara State at a Latitude of $N8^014^I45^{II}$ and Longitude $E4^048^I42^{II}$. This material was collected directed from the factory furnace at a calcination temperature of 700°C. Ordinary Portland Cement (OPC), fine aggregate and coarse aggregate were sourced in Sango Ota, Ogun State . Calcium Nitrate(CN) was purchased from a local vendor in Osogbo, Osun State. Portable water was obtained at the Building Research Laboratory of the Nigerian Building and Road Research Institute (NBRRI), Ota, Ogun State where the experiments were performed.

Testing of materials
Characterization of OPC, RHA and CN were examined using X-ray fluorescence analysis which was carried out at the Petroleum Reservoir Laboratory under the department of Chemical and Petroleum Engineering Afe Babalola University, Ado-Ekiti, Ekiti. The oxides composition such as silica (SiO_2), alumina (Al_2O_3), ferric oxide (Fe_2O_3), calcium oxide (CaO), magnesium oxide (MgO), Sulphur trioxide (SO_3) etc. were determined in line with [23, 25]. The physical properties of RHA, river sand (fine aggregate) and granite (coarse aggregate) determined were particle size distribution and specific gravity in line with the existing standard.

Concrete mix proportion
The concrete mix ratio of 1:2:4 and water-cement ratio of 0.6 was adopted in the study, which is similar to [20]. As supported by the findings of [12, 22, 27], the optimum RHA content in blended concrete to enhance strength and workability is 15%. This study adopted a fixed RHA content of 15% as presented in Table 1. Sample without RHA and CN represent control 1 while control 2 represent sample with only RHA content. The remaining samples contained 15% content of RHA with varying percentage of CN from 1 to 5.

Advances in Cement and Concrete   Materials Research Forum LLC
Materials Research Proceedings 51 (2025) 29-38   https://doi.org/10.21741/9781644903537-4

Table 1. Batching of material for production of concrete specimen

Samples Identification	Specimen	(kg)	OPC (kg)	RHA (kg)	CN (kg)	FA (kg)	CA
A	100% OPC (Control 1)	50	0	0	100	200	
B	85% OPC+15% RHA (Control 2)	42.5	7.5	0		100	200
C	84% OPC + 15% RHA + 1%CN	42	7.5	0.5	100	200	
D	83% OPC + 15% RHA + 2%CN	41.5	7.5	1.0	100	200	
E	82% OPC + 15% RHA + 3%CN	41	7.5	1.5		100	200
F	81% OPC + 15% RHA + 4%CN	40.5	7.5	2.0	100	200	
G	80% OPC + 15% RHA + 5%CN	40	7.5	2.5	100	200	

OPC= Ordinary Portland Cement   RHA= Rice Husk Ash   CN
$(Ca(NO_3)_2)$ = Calcium Nitrate   FA = Fine Aggregate   CA = Coarse Aggregate

Results and Discussion
The results and discussion of the research were discussed in this section:
Characterization of materials
The chemical composition of OPC, RHA and CN is presented in Table 2. From the result of XRF analysis, it was noticed that RHA belongs to Class F pozzolan because the sum of Silica (SiO_2), Alumina (Al_2O_3) and Ferric (Fe_2O_3) is 75.36% which is more than 70% as specified according to [6] and with LOI result of 1.5% which is less than 6% as specified by [3] for class F pozzolans. Among the XRF results of RHA determined by authors like [22, 24, 1, 2] were in agreement with this study. Also, XRF result of CN and OPC shows that CaO values are 81.95 and 68.76%, respectively which is higher than that of RHA therefore make it more cementitious material. Rice husk ash (RHA) satisfies the maximum value of 5% specified by [6] for the SO_3 of cementitious material with a value of 0.5%.

Table 2. Chemical composition of RHA, CN and OPC

Chemical Constituent	Percentage Composition (%)		
	RHA	CN	OPC
SiO_2	69.32	2.05	18.65
Al_2O_3	3.41	0.60	3.24
Fe_2O_3	2.63	5.01	3.24
CaO	5.62	81.95	68.76
MgO	0.73	5.61	1.82
Na_2O_3	0.32	0.43	0.37
K_2O	6.16	1.10	1.65
P_2O_5	0.61	0.05	-
SO_3	0.50	1.60	2.90
LOI	1.50	1.49	1.73

Physical properties of RHA, sand and coarse aggregate

This section outlines the results of the laboratory investigation into the physical properties of RHA, sand, and aggregate, covering aspects such as particle size distribution and specific gravity.

Particle size distribution

The results of the particle size distribution (PSD) analysis carried out in this research work were graphically presented in Figure 1. The grading curve of aggregates shows the coefficient of uniformity (Cu) and coefficient of curvature (Cc) values were 3.3 and 1.1 for river sand, 2.0 and 1.09 for granite, respectively. The Cu value for sand is less than 4 and Cc value is between 1 to 3 as specified by unified soil classification system (USCS) as postulated in [7]. Thus, it can be categorized as well graded sand and gravelly sands. The Cu gives an indication of the spread of particle size value [21]. The Cu for granite that is 1.0 can be classified as uniformly graded. RHA used was sieve through sieve No 200. Based on these assessments, river sand, granite and RHA are suitable ingredients for making quality concrete [8].

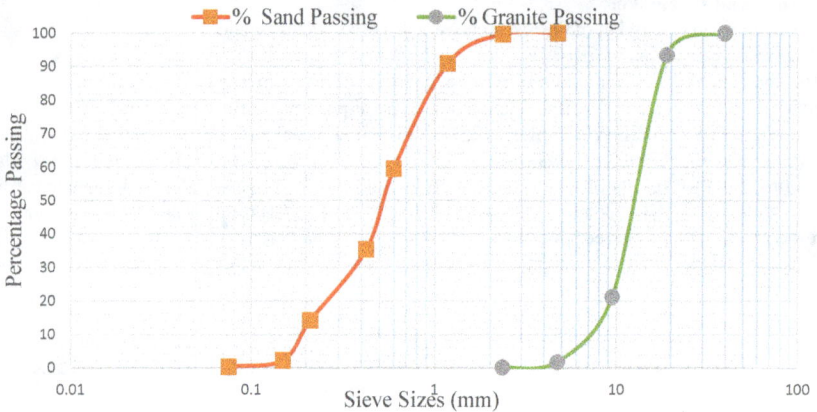

Fig. 1. Particle size analysis of sand and granite

Specific gravity

The results of specific gravity of the materials used in this study were determined as presented in Table 3. The specific gravity value of RHA is lower than that of river sand and granite. Therefore, RHA is expected to produce a less dense concrete with better thermal, insulative properties and significant reduction in dead weight than the conventional aggregate when use in concrete [29]. The specific gravity value of RHA (2.27 g/cm³) fall outside the range of 2.5-3.0 g/cm³ specified by [10] for normal weight aggregate and can be regarded as low weight aggregate. The specific gravity of RHA (2.27 g/cm³) is in correlation with the result of [14]. The specific gravity values obtained for granite (2.71 g/cm³) and river sand (2.62 g/cm³) fall within the range of 2.5-3.0 g/cm³ as specified by [17]. It is also in accordance with [5] standard specification for concrete aggregates which acknowledges that normal-weight of aggregates commonly has specific gravities between 2.4 g/cm³ and 2.9 g/cm³.

Advances in Cement and Concrete
Materials Research Proceedings 51 (2025) 29-38

Materials Research Forum LLC
https://doi.org/10.21741/9781644903537-4

Table 3. Specific gravity of RHA, OPC, sand and granite

Materials	Specific Gravity
RHA	2.27
OPC	3.10
Fine Aggregate (Sand)	2.62
Coarse Aggregate (Granite)	2.71

Workability
This section discusses the workability (slump and compacting factor) and setting time of the control samples and varying RHA-CN blended cement.

Slump
The result of the slump test is presented in Figure 2. This shows that inclusion of RHA and CN as substitute of cement reduced collapse. It shows that sample A and B concrete had slump value of 39 and 36 mm respectively while the RHA-CN mixes at intervals of 1 to 5% are 34, 32, 26, 19 and 14 mm respectively. This shows that the slump values decreased from 39 to 36 mm after the addition of 15% RHA for sample B while later reduced from 34 to 14 mm as the percentage of CN content increases from 1 to 5%. Therefore, as the percentage of CN content increases, the slump values reduces. In order to attain the required workability, mixes containing RHA-CN will require more water than conventional mixtures. This is similar to the report of [18], that workability of the concrete decreases as the percentage of CN increases. The addition of CN increases the formation of calcium-related compounds, such as calcium silicate hydrates and ettringite. These compounds fill the voids between blended cement particles and aggregates, making the mix more compact and less able to flow, leading to a lower slump. The decrease in slump values with higher CN content is likely attributed to accelerated hydration, quicker consumption of free water, and an increase in mix stiffness. This leads to reduced workability, which is reflected in lower slump values.

The slump test results suggest that while RHA-CN blended concrete has reduced workability due to the formation of calcium-related compounds, it offers significant benefits in terms of strength, durability, and environmental sustainability. Its reduced workability can be mitigated by using appropriate admixtures and mechanical compaction techniques, making it suitable for various civil engineering applications, particularly in projects requiring high strength and durability [23]. Additionally, [26] concluded that the use of RHA in blended cement concrete improves durability while maintaining acceptable mechanical properties. Therefore, careful mix design and construction planning are essential to harness the full potential of RHA-CN concrete in modern construction [17].

Materials Research Forum LLC

https://doi.org/10.21741/9781644903537-4

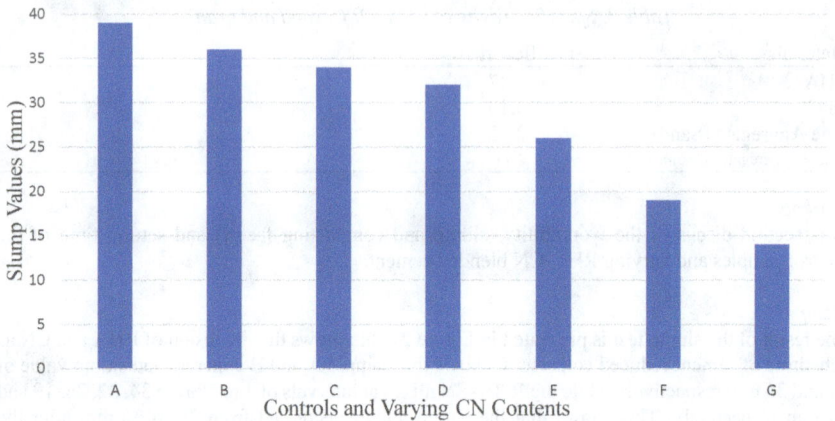

Fig. 2. Slump of RHA-CN concrete

Compacting factor

The result of compacting factor as presented in Figure 3 shows that specimen A and B have value of 0.92 and 0.89, respectively. The specimens with varying CN content of 1, 2, 3, 4 and 5 % have values of 0.84, 0.81, 0.77, 0.75 and 0.72 respectively. This shows that compacting factor decreases as the CN content increases. This is supported with the result of [16] that categorized compacting factor values as very low (0.78), low (0.85), medium (0.92) and high (0.95) in accordance with building research establishment. This result shows that sample A and B category fall in medium degree workability. Sample C and D fall in low degree while sample E, F and G were very low degree of workability. As the addition of CN increases the concrete workability becomes low.

The implication of this result is that medium degree workability (Sample A and B) mixes will be relatively easy to place and compact, making them suitable for most general construction projects. They can be used where manual compaction or light mechanical vibration is available, such as in slabs, foundations, and beams. Sample C and D (blended concrete with low workability) requires more effort to place and compact, which could lead to improper consolidation if sufficient vibration or compaction is not applied. This type of mix is typically used in pavement construction or precast concrete elements, where formwork or mechanized compaction is available.

For Sample E, F, and G (very low degree workability), these mixes are difficult to handle and will require intense mechanical compaction, such as vibration, to achieve full consolidation. If not compacted properly, these mixes can lead to the formation of voids, reducing the strength and durability of the concrete. Such mixes are suitable for applications like mass concrete or where low water-cement ratios are required for strength purposes, but adequate compaction methods must be ensured. In addition, very low workability mixes are often used in high-strength or specialized concrete, where water-cement ratios are minimized to enhance strength and durability. For these cases, the low workability is acceptable as long as proper compaction and vibration are applied. This is common in prestressed concrete and high-performance concrete applications, where strength is prioritized over workability.

Advances in Cement and Concrete   Materials Research Forum LLC
Materials Research Proceedings 51 (2025) 29-38   https://doi.org/10.21741/9781644903537-4

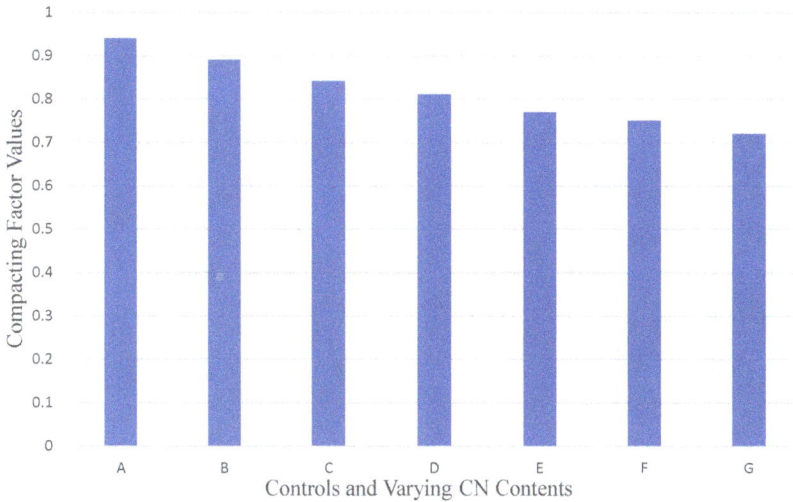

Fig. 3. Compacting factor of RHA-CN concrete

Setting time

The variation in setting time for A, B and varying percentage of CN were shown in Figure 4. It was observed that the initial setting time and final setting time are within the provision of [9]. The initial and final setting time increased from A to B. The addition of CN reduces the initial and final setting time from C to G. This may be attributed to the increase in the concentration of calcium cations which boosted the low calcium oxide content of RHA. This agrees with the report of [18]. This implies that the setting time is sensitive to CN content. As the percentage of CN content kept increasing, the hydration process is high resulting in a lower setting time. The result of sample B is in accordance with the work of [22] which concludes that the initial setting time for RHA blended cement is longer than that of sample A. The minimum initial setting time requirement (not less than 30 minutes) was met by the blended cement at all replacement levels considered.

The setting time data could be used by ready-mix concrete suppliers to determine the appropriate transportation and placement time for concrete mixes. This is to ensure that the concrete arrives at the construction site with sufficient workability and does not set prematurely. This is especially relevant in projects with long transportation distances or complex logistics, such as urban infrastructure projects, airport runways, and tunnels, where concrete must remain workable for an extended period before placement.

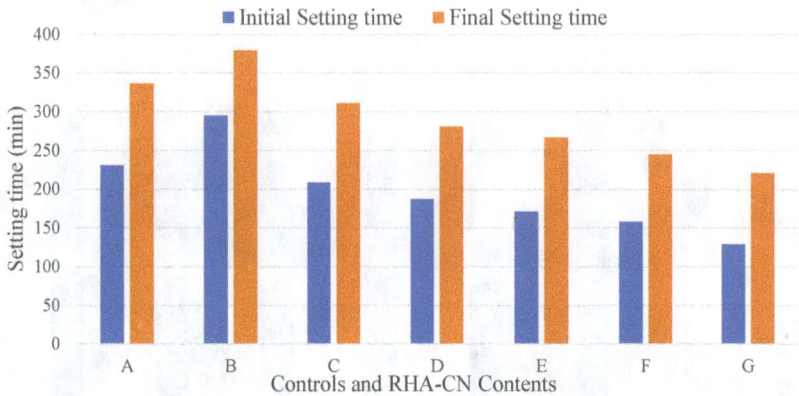

Fig. 4. Initial and final setting time of RHA-CN concrete

Conclusions

The following conclusions were drawn from this research work:

i. RHA is suitable for use as mineral admixture in concrete as the addition of SiO_3, Al_2O_3 and Fe_2O_3 contents present were 75.36% which had satisfied the condition of a pozzolanic material;

ii. the addition of CN has improved the workability and setting time of fresh concrete made with RHA-CN and

iii. concrete containing 5% CN addition was found to be the optimum content.

Acknowledgements

The author(s) would like to thank the Nigerian Building and Road Research Institute for allowing the use of their facilities to conduct this research.

References

[1] Ahmed, A., Hyndman, F., Kamau, J. and Fitriani, H. Rice husk ash as a cement replacement in high strength sustainable concrete. Mater Sci Forum. 10,7 (2020) 90–98.

[2] Ali, T., Saand, A., Bangwar, D., Buller, A., and Ahmed, Z. Mechanical and durability properties of aerated concrete incorporating rice husk ash (RHA) as partial replacement of cement. Crystals.11 (2021) 604. https://doi.org/10.3390/cryst11060604

[3] American Society for Testing and Materials. Standard specification for coal fly ash and raw or calcined natural pozzolan for use in concrete. ASTM C618 (2021). Retrieved from: https://www.astm.org/Standards/C618

[4] Arum, C., Ikumapayi, C. M. and Aralepo, G. O. Ashes of biogenic wastes: Pozzolanicity, prospects for use, and effects on some engineering properties of concrete. (2013). Retrieved from: https://scite.ai/reports/10.4236/msa

[5] ASTM C33/C33M. Standard specification for concrete aggregates. ASTM International; (2023).

Advances in Cement and Concrete
Materials Research Proceedings 51 (2025) 29-38

Materials Research Forum LLC
https://doi.org/10.21741/9781644903537-4

[6] ASTM C618-12. Standard specification for coal fly ash and raw or calcined natural pozzolan for use in concrete. ASTM International, West Conshohocken, PA; (2012). https://doi.org/10.1520/C0618

[7] ASTM D2487. Standard practice for classification of soils for engineering purposes. ASTM International, West Conshohocken, PA; (2011).

[8] Bowles, J. E. Engineering properties of soils and their measurements. 4th ed. New York: McGraw-Hill (1992).

[9] British Standards Institution. Methods of testing cement: Determination of setting times and soundness. BS EN 196: Part 3. London: BSI (2005).

[10] BS EN ISO 10545. Ceramic tiles. Determination of water absorption, apparent porosity, apparent relative density, and bulk-density. British Standards Institution, London (1997).

[11] Faria, K. C. P., Gurgel, R. F. and Holanda, J. N. F. Recycling of sugarcane bagasse ash waste in the production of clay bricks. J Environ Manage. 101 (2012) 7-12. https://doi.org/10.1016/j.jenvman

[12] Ganesan, K., Rajagopal, K. and Thangavel, K. Rice husk ash blended cement: Assessment of optimal level of replacement for strength and permeability properties of concrete. Constr Build Mater.22,8 (2008) 1675-1683. https://doi.org/10.1016/j.conbuildmat

[13] Givi, A. N., Rashid, S. A., Aziz, F. N. A. and Salleh, M. A. M. Contribution of rice husk ash to the properties of mortar and concrete: A review. Constr Build Mater. 24,11 (2010) 1231-1245. https://doi.org/10.1016/j.conbuildmat

[14] Kazmi, S. M. S, Abbas, S., Munir, M. J, and Khitab, A. Exploratory study on the effect of waste rice husk and sugarcane bagasse ashes in burnt clay bricks. J Build Eng. 7 (2016) 372-378. https://doi.org/10.1016/j.jobe

[15] Nair, D.G, Jagadish, K. S, Fraaij, ALA. Reactive pozzolanas from rice husk ash: An alternative to cement for rural housing. Cem Concr Res. 36,6 (2006) 1062-1071. https://doi.org/10.1016/j.cemconres

[16] Neville, A. M and Brooks, J. J. Concrete technology. Pearson Education; (2010).

[17] Neville, A. M. Properties of concrete. 5th ed. Pearson Education Limited; (2011).

[18] Ogunbode, E. B. and Hassan, I. O. Effect of addition of calcium nitrate on selected properties of concrete containing volcanic ash. Leonardo Electron J Pract Technol. 19 (2011) 29-38.

[19] Okonkwo, C. E. and Igwe, C. C. The potential of rice husk ash as a supplementary cementitious material in Nigeria: A review. J Constr Build Mater. 46 (2021) 110-122.

[20] Oriola K. O., Raheem A. A., Kareem, M. A. and Abdulwahab, R. Assessment of workability and compressive strength of rice husk ash-blended palm kernel shell. J Civil Environ Stud. 7,1 (2021). https://doi.org/10.36108/

[21] Raheem, A. A. and Ikotun, B. D. Incorporation of agricultural residues as partial substitution for cement in concrete and mortar: A review. J Build Eng. 31 (2020). https://10.1016/j.101428

[22] Raheem, A. A. and Kareem, M. A. Optimal raw material mix for the production of rice husk ash blended cement. Int J Sustain Constr Eng Technol. 7,2 (2017a).

[23] Raheem, A. A. and Kareem, M. A. Chemical composition and physical characteristics of rice husk ash blended cement. Int J Eng Res Afr. 32 (2017b) 25–35.

[24] Raisi, E. M, Amiri, J. V, Davoodi, M. R. Mechanical performance of self-compacting concrete incorporating rice husk ash. Constr Build Mater. 177 (2018) 148–157. . https://doi.org/10.1016/j.conbuildmat

[25] Seyed, A. Z., Farshad, A., Farzan, D. and Mojtaba A. Rice husk ash as a partial replacement of cement in high strength concrete containing micro silica. Case Stud Constr Mater. 7 (2017) 73-81. https://doi.org/10.1016/j.cscm

[26] Shafigh, P., Mahmud, H. B. and Jumaat, M. Z. Effect of calcium nitrate on workability and early-age strength of high-performance concrete. Mater Des. 32,5 (2011) 3187-3193. https://doi.org/10.1016/j.matdes

[27] Swetha, R., Rani, M. S. and Raju, U. Experimental works on self-compacting concrete by partial replacement of rice husk ash with subjected to acid attack. Int J Innov Technol Explor Eng. 9 (2020) 664–667. https://doi.org/10.35940/ijitee

[28] Tayeh, B. A., Alyousef, R., Alabduljabbar, H. and Alaskar, A. Recycling of rice husk waste for a sustainable concrete: A critical review. J Clean Prod. 312 (2021) 127-134. https://doi.org/10.1016/j.jclepro

[29] Teo, D. C. L., Mannan, M. A., Kurian, V. J. and Ganapathy, C. Lightweight concrete made from oil palm shell (OPS): Structural bond and durability properties. Build Environ. 42 (2007) 2614–2621. https://doi.org/10.1016/j.buildenv

[30] Thomas, M. D. A., Shehata, M. H. and Shashiprakash, S. G. Use of supplementary cementing materials in concrete: Basics, benefits, and the influence of calcium nitrate on workability. ACI Mater J. 109,6 (2012) 767-776.

[31] Zhang, M. H., Islam, J. and Peethamparan, S. Use of calcium nitrate as an accelerator in combination with supplementary cementitious materials. Cem Concr Compos. 60 (2015) 50-59. https://doi.org/10.1016/j.cemconcomp

Advances in Cement and Concrete
Materials Research Proceedings 51 (2025) 39-48

Materials Research Forum LLC
https://doi.org/10.21741/9781644903537-5

The influence of wood ash fineness on early strength development of hollow sandcrete blocks

Akeem Ayinde RAHEEM[1,a], Favour Wuraola KOLAWOLE[2,b*],
Isaiah Oluwafemi OGUNTOLA[3,c], Oludayo Ajani AKINTOLA[2,d]

[1]Department of Building, Ladoke Akintola University of Technology, Ogbomoso, Nigeria

[2]Department of Civil Engineering, Ladoke Akintola University of Technology, Ogbomoso, Nigeria

[3]Nigerian Building and Road Research Institute, Km10, idiroko road, Ota, Ogun State, Nigeria

[a]aaraheem@lautech.edu.ng [b]kolawoleoludara@gmail.com [c]oguntolaisaiah@nbrri.gov,
[d]oludayo.akintola@gmail.com

Keywords: Wood Ash, Pozzolan, Compressive Strength, Sandcrete Block, Water Absorption

Abstract. This study investigates the use of Wood Ash (WA) as partial replacement for ordinary Portland cement (OPC) in sandcrete blocks. Hollow sandcrete blocks were produced by partially replacing the cement content with 15% of WA by weight in different fineness, (425 ,250 ,125 ,75 μm), Blocks without WA serve as the control. The blocks were produced with the use of Vibrating block moulding machine with 6" (450mm x 225mmx150mm) mould. Mix ratio (1:8) cement-sharp sand ratio was used as recommended by NIS 87:2007. The blocks were tested for compressive strength, density, and water absorption at 7,14, 28 days. The results showed that the wood ash belongs to Class C pozzolan due to its high CaO content of 55.90%, which is well above the 20% typically required for Class C pozzolans, as specified in [6]. The compressive strength of sandcrete hollow blocks at 28 days were 2.14N/mm²(Control), 1.078N/mm²,1.088N/mm²,1.101N/mm², and 1.252N/mm² for blocks with 15% wood ash fineness 425μm, 250μm, 125μm, 75μm, respectively. The density obtained at 28 days were 1884.25 kg/m³(Control), 1520.74 kg/m³, 1597.51 kg/m³,1632.59 kg/m³ and 1741.06 kg/m³ for blocks with 15% wood ash fineness 425μm, 250μm, 125μm, 75μm, respectively. The water absorption obtained at 28 days are 8.84%(Control), 19.35%, 14.35% ,7.319% and 4.172%. for blocks with 15% wood ash fineness 425μm, 250μm, 125μm, 75μm, respectively. Blocks with finer wood ash 75 μm showed better compressive strength, density, and lower water absorption compared to coarser ash. The use of wood ash improved the durability characteristics of sandcrete blocks.

Introduction

The need for affordable building materials has become increasingly important in addressing the housing demands of the world's growing population. As the cost of conventional materials continues to increase, many people find themselves unable to afford proper housing, especially in areas where poverty is widespread [12]. This has led to a growing interest in exploring alternative, locally available materials that can be used to construct functional, low-cost buildings in both rural and urban area. Examples of such materials include earthen plaster [30], lateritic interlocking blocks [25]. Meanwhile, the continuous accumulation of waste from industrial by-products and agricultural residues poses significant environmental challenges, particularly regarding their treatment and disposal. A survey by the Raw Materials Research and Development Council of Nigeria highlights several local building materials that hold promise as substitutes for imported ones [5]. Among these materials are cement/lime-stabilized bricks or blocks, sundried (adobe) soil blocks, burnt clay bricks, cast in-situ walls, rice husk ash (RHA), wood ash (WA), mud and straw, lime, and stonecrete blocks. One way to reduce cement usage is by partially replacing it with a

Advances in Cement and Concrete   Materials Research Forum LLC
Materials Research Proceedings 51 (2025) 39-48   https://doi.org/10.21741/9781644903537-5

pozzolan—a siliceous material that has little cementitious value on its own but, when combined with lime in the presence of water, enhances the durability of cement. Various studies have explored the use of pozzolans as partial replacements for cement in producing hollow sandcrete blocks [32, 23, 31, 18, 17, 28]. [10] studied the use of wood waste ash as a partial cement replacement in structural concrete and mortar. WA, a byproduct of wood combustion, is known for its pozzolanic properties. Several researchers have explored the partial replacement of cement with wood ash in concrete [1, 27 ,21, 11]. [29] determined that an optimal dosage of 15% wood ash can achieve the desired compressive strength for concrete blocks. In this research, the same 15% wood ash replacement was investigated, focusing on varying degrees of fineness to assess its effect on the compressive strength of hollow sandcrete blocks.

[21] used Wood Ash from a bread bakery at Ladoke Akintola University of Technology, Ogbomoso to replace 5% - 25% by weight of the cement in concrete. 1:2:4 mix ratio was used with water to binder ratio maintained at 0.5. The compression strength of specimens was determined at curing ages 3, 7, 28, 56, 90, and 120 days and it was concluded that only up to 10% wood ash replacement is suitable for structural concrete.

The present work evaluates the potential of wood ash fineness in enhancing the properties of sandcrete hollow blocks. compressive strength, density and water absorption of wood ash

Materials and Methods
The main materials used for this research include cement, fine aggregate, clean water and Wood ash.

Cement
The cement used for this experiment was Ordinary Portland Cement (OPC) of grade 32.5R (3X Dangote Cement Brand) which conformed to [15]. This standard specifies the quality requirements for cement used in building and construction to ensure adequate strength, durability and consistency.

Fine aggregate
The fine aggregate was sourced from a local supplier in Ogbomoso, Nigeria. It was free of silt and debris, which can hinder the strength development of the blocks.

Water
The water used for the production of blocks was clean, colorless, odorless, free from any type of organic impurities and was obtained from LAUTECH Building Technology Laboratory.

Wood ash
The wood ash sample that was used in this experiment was sourced from Ladoke Akintola University of Technology Ogbomoso, a bread bakery situated at the back of the Food Science and Engineering Technology. It was carefully gathered from the bakery's exhaust using a hand scoop after the fuel had completely burned out.

Methods

Chemical Composition of Wood Ash (WA)
The chemical analysis of WA to determine its Pozzolanicity was carried out in the laboratory of Covenant University, Ota, Ogun State, Nigeria. The main oxide composition of the WA (SiO_2, Al_2O_3, Fe_2O_3, CaO) was determined. Minor oxide such as Na_2O, MgO, K_2O.... Etc. was also examined and compared with the requirements for a good pozzolan. The result is shown in Table 2.

Advances in Cement and Concrete
Materials Research Proceedings 51 (2025) 39-48

Materials Research Forum LLC
https://doi.org/10.21741/9781644903537-5

Grading of Fine aggregates
Sieve analysis of the sand was used to determine the grading of the soil sample in accordance with procedure in [9].

Production of Blocks
The blocks were produced using a vibrating block moulding machine with a 6" (450 mm x 225 mm x 150 mm) mould. A cement-to-sand mix ratio of 1:8 was adopted, as recommended by [17]. Cement was partially replaced with 15% wood ash. The wood ash was passed through specific sieve sizes 425 μm, 250 μm, 125 μm, and 75 μm. It was ensured that only the particles smaller than each size was collected. Hand mixing was employed, and the materials were thoroughly turned multiple times until a uniform color and consistency were achieved. Water was added gradually to ensure the mixture was workable, and the correct amount was judged by pressing a sample between the palms—if it caked without releasing water, it was considered sufficient [24]. The composite mixture was then placed into the mould in the block moulding machine and vibrated for one minute to ensure proper compaction, as recommended by [24]. The freshly molded block, resting on a wooden pallet, was removed from the machine and placed on the ground for curing. Water was sprinkled on the blocks at least twice a day to ensure proper curing over a period of twenty-eight days. Figure 1 shows the vibrating block moulding machine and table 1 shows the mix proportion for the blocks.

Curing of the specimens
The block specimens were stored in a place free from vibration, not expose to direct sunlight, and it was subjected to curing. The curing period span 7, 14, and 28 days respectively. The curing was done by wetting the blocks with water twice in a day morning and afternoon.

Testing of the blocks
The following tests were conducted on Hollow Sandcrete Blocks: The compressive strength, Density and water absorption

Compressive strength
Compressive strength test was carried out to determine the load bearing capacity of the blocks. The blocks that have attained the ripe ages for compressive strength test of 7, 14, 28, days was taken from the curing area to the laboratory, two hours before conducting the test at Kwara State Ministry of Works laboratory for test, to normalize the temperature and to make the block free from moisture.

Table 1: Mix proportions for WA blocks

Specimen	Mix proportion (kg)		
	Cement	**Wood ash**	**Fine Aggregate**
Control (100%OPC)	50	0	400
85%OPC+15%WA(425μm)	42.5	7.5	400
85%OPC+15%WA(250μm)	42.5	7.5	400
85%OPC+15%WA(125μm)	42.5	7.5	400
85%OPC+15%WA(75μm)	42.5	7.5	400

The weight of each block was taken before being placed on the compression testing machine in between metal plates. The blocks were crushed and their corresponding failure loads were recorded. The crushing force was divided by the gross sectional area of the block to give the compressive strength. The strength value was the average of three specimens.

$$Compressive\ strength\ (MPa)\ F = \frac{Load\ (N)\quad P}{Cross-sectional\ Area\ (mm^2)\quad A} \qquad 1$$

Advances in Cement and Concrete
Materials Research Proceedings 51 (2025) 39-48

Materials Research Forum LLC
https://doi.org/10.21741/9781644903537-5

Density
The density of the blocks was determined by dividing the weight of the block prior to crushing, with the net volume. The density value was the average of three specimens.

$$D = \frac{mass\ of\ block(kg)}{Dimensional\ volume\ of\ block\ (m^3)} \qquad\qquad 2$$

Water absorption
This was performed on the sandcrete blocks after curing for 28 days. Three samples per WA content were removed from the curing area and sun dried until there is no further loss in their dry weights. The samples were then immersed in water for 24hours and allowed to drain for 10 minutes before taking their wet weights. The difference in weight was used to calculate the percentage water absorbed for each block as follows:

$$Water\ Absorption(\%) = \left[\frac{M2-M1}{M1}\right] \times 100 \qquad\qquad 3$$

M_2 = Weight of wet block
M_1 = Weight of dried block

Results and Discussion
The results obtained from the various tests performed are discussed in the following sections.

Chemical composition of WA
The chemical composition of wood ash is shown in Table 2. It was observed that the wood ash belongs to Class C pozzolan due to its high CaO content of 55.90%, which is well above the 20% typically required for Class C pozzolans, as specified in [6]. The sum of Silica (SiO_2), Alumina (Al_2O_3), and Ferric Oxide (Fe_2O_3) is 32.95%, which is lower than the 50% threshold often used to classify pozzolans as Class C.

Table 2: Chemical composition of Wood ash

Chemical composition	Percentage Composition
Silica (SIO_2)	10.20
Alumina (AL_2O_3)	20.50
Iron (FE_2O_3)	2.25
Calcium (CaO)	55.90
Magnesium oxide (MgO)	4.10
MnO	0.05
Sodium (Na_2O)	1.70
Potassium (K_2O)	0.60
Ni	21.20
Loss on ignition (LOI)	0.04

Sieve Analysis of Fine Aggregates
The grading of sharp sand used in the production of hollow sandcrete blocks is as showed in Figure 2 It could be observed from the figure that the coefficient of uniformity (Cu) and coefficient of curvature (Cc) for the sharp sand are 4.13, 1.12 and 4.15, 1.30. The distribution of particle sizes is shown by the coefficient of uniformity [19]. The fine aggregates is uniformly graded and can be used to produce high-quality blocks because the coefficient of uniformity values fall between 1.0 and 5.0. The sharp sand meets the British Standard requirements for fine aggregates under the fine grading zone as specified in [9] and therefore suitable for use in the production of sandcrete block

Advances in Cement and Concrete   Materials Research Forum LLC
Materials Research Proceedings 51 (2025) 39-48   https://doi.org/10.21741/9781644903537-5

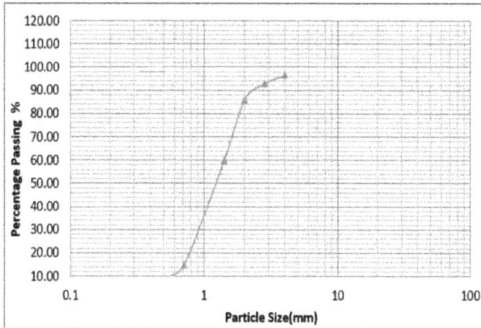

Figure 2 *Particle Size Distribution graph of Fine Aggregate*

Compressive Strength

The compressive strength results of sandcrete hollow blocks are presented in Figure 3. The compressive strength of sandcrete blocks made with 100% OPC (the control) at 7,14,28 days were 1.43 N/mm^2, 1.58 N/mm^2, 1.72 N/mm^2 from 7 to 28 days. Which falls below the minimum 28 days compressive strength of 2.50N/mm^2 stipulated by Nigerian Industrial Standard [14].

The compressive strength of sandcrete hollow blocks are presented in figure 3. The compressive strength of sandcrete blocks made with 15%WA (425µm) at 7, 14, 28 days were 0.819 N/mm^2,0.89N/mm^2,1.078N/mm^2 respectively. These results are lower than that obtained by [24] with values ranging from 1.01 N/mm^2 to 1.68 N/mm^2, during the same period. This may be due to differences in the physical properties of the sand used which depend on the source. However, the value was below minimum 28 days compressive strength of 2.50N/mm^2 stipulated by Nigerian Industrial Standard [14]

The compressive strength of sandcrete blocks made with 15%WA (250µm) at 7,14,28 days were 0.887N/mm^2,0.953 N/mm^2,1.088 N/mm^2 respectively. These results are lower than those obtained by [22] who worked on the effect of replacing cement with 15% sawdust ash, their compressive strength was reported to be 1.6N/mm^2 at 28 days. This may be due to differences in the chemical properties of the WA used which depend on the source. However, the value falls below the minimum 28 days compressive strength of 2.50N/mm^2 stipulated by Nigerian Industrial Standard [14]

Sandcrete blocks made with 15%WA (125µm) at 7,14,28 days were 0.895 N/mm^2, 0.99 N/mm^2, 1.101 N/mm^2 respectively. These results are lower than that obtained by [2] who worked on the properties of sandcrete blocks with sawdust ash replacement, the compressive strength was around 1.23N/mm^2 at 28days. This may be due to differences in the physical properties of the sand used which depend on the source. However, the value falls below the minimum 28 days compressive strength of 2.50N/mm$_2$ stipulated by Nigerian Industrial Standard [14]

The compressive strength of sandcrete blocks made 15%WA (75 µm) were 1.015 N/mm^2,1.078 N/mm^2,1.252 N/mm^2 When compared to [24] These results are lower than those obtained by [24] with values ranging from 1.01 N/mm^2 to 1.68 N/mm^2, during the same period.

Advances in Cement and Concrete   Materials Research Forum LLC
Materials Research Proceedings 51 (2025) 39-48   https://doi.org/10.21741/9781644903537-5

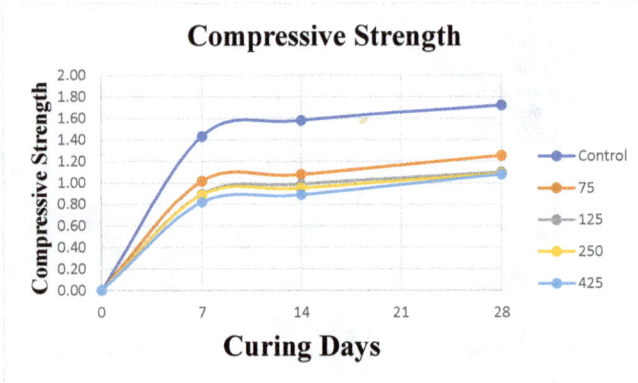

Figure 3 *Effect of percentage replacement of WA fineness on compressive strength of Sandcrete Blocks*

Density

The Density results are presented in Figure 4, the results show blocks without WA. The density ranges from 1495.34 kg/m³ to 1584.25kg/m³ the densities were above the minimum value of 1500kg/m3 recommended for first grade sandcrete blocks by [13].

15%WA (425µm) WA replacement, it was observed that there was decrease in density as the curing age increases. The density obtained from the blocks; 7-28 days ranged from 1509.27 kg/m³ to 1520.74 kg/m³. These values are lower than that of [2] which ranged from 1573.32 kg/m³ to 1597.51 kg/m³ The densities were above the minimum value of 1500kg/m³ recommended for first grade sandcrete blocks by [13].

For 15%WA (250µm) WA replacement as shown on Figure 4, it was observed that there was decrease in density as the curing age increases. The density obtained from the blocks, 7-28 days ranged from 1573.32 kg/m³ to 1597.51 kg/m³. These values are slightly close to that of [21]. which ranged from 1573 kg/m³ to 1597 kg/m³. The densities were above the minimum value of 1500kg/m³ recommended for first grade sandcrete blocks by [13].

The result of 15%WA (125µm) WA replacement as shown in Figure 4, it was observed that there was decrease in density as the curing age increases. The density obtained from the blocks, 7-28 days ranged from 1592.27 kg/m³ to 1632.59 kg/m³. These values are higher than that [3] which ranged from 1550 kg/m³ to 1580 kg/m³. The densities were above the minimum value of 1500kg/m³ recommended for first grade sandcrete blocks by [13].

For 15%WA (75µm) WA replacement as shown on Figure 4, it was observed that there was decrease in density as the curing age increases. The density obtained from the blocks, 7-28 days ranged from 1673.12 kg/m³ to 1741.06 kg/m³. These values are higher than that of [2] which ranged from 1573.32 kg/m³ to 1597.51 kg/m³ The densities were above the minimum value of 1500kg/m³ recommended for first grade sandcrete blocks by [13].

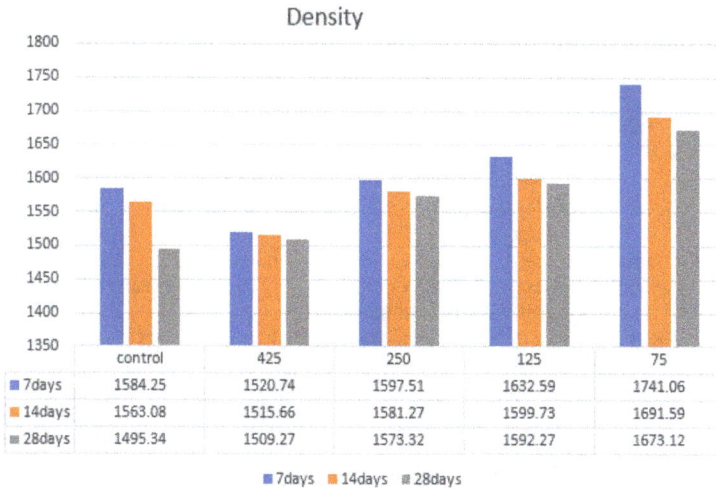

	control	425	250	125	75
7days	1584.25	1520.74	1597.51	1632.59	1741.06
14days	1563.08	1515.66	1581.27	1599.73	1691.59
28days	1495.34	1509.27	1573.32	1592.27	1673.12

Figure 4 Effect of percentage replacement of WA fineness on Density of Sandcrete Blocks

Water Absorption

Figure 5 shows the variation of water absorption of Sample A, B, C, D, and E, at day 28.

1.64%, 3.47%, 2.62%, 1.33%, 0.766%. 15% wood ash 425 µm was 3.47%. This means wood ash with higher fineness absorb more water due to particle sizes and more voids allowing more water to absorb, compare to 15% wood ash 75µm which is 0.766% Sandcrete blocks with 15%WA 425 µm replacement is the most porous with the absorption rate of 19.35%. This absorption rate is higher than 16.95% obtained by [4] and the acceptable value of 12% according to [8]

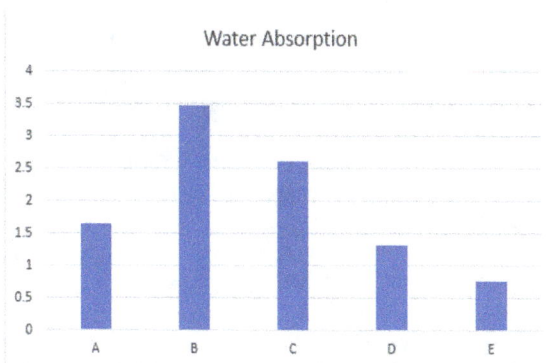

Figure 5 Effect of percentage replacement of WA fineness on Water Absorption on Sandcrete Blocks

Advances in Cement and Concrete
Materials Research Proceedings 51 (2025) 39-48

Materials Research Forum LLC
https://doi.org/10.21741/9781644903537-5

Keys:

A - Control (100%OPC)
B - 85 OPC + 15%WA (425µm)
C - 85 OPC + 15%WA (250µm)
D - 85 OPC + 15%WA (125µm)
E - 85 OPC + 15%WA (75µm)

Conclusions

The following conclusions are drawn from the present study;

i. WA is suitable for use as mineral admixture in sandcrete block. It is Classified as a Class C pozzolan according to [6].

ii. Compressive strength, density and water absorption results indicate that the different fineness levels of wood ash (425 ,250 ,125 ,75 µm) significantly affect the strength and performance of sandcrete blocks

iii. The compressive strength of sandcrete blocks increases as the curing age increases but decreases as the WA Fineness content increases. The finer the WA the higher the strength

iv. Blocks with finer wood ash 75 µm showed better compressive strength, density, and lower water absorption compared to coarser ash. The use of wood ash improved the durability characteristics of sandcrete blocks. To obtain higher strength adequate for use in sandcrete hollow blocks for non-load bearing walls in buildings, this research suggests that extending the curing period beyond 28 days can further enhance the compressive strength and overall durability of the blocks. Future studies should explore the benefits of extended curing times, such as 56 or 90 days, to determine the optimal duration for blocks containing wood ash.

References

[1] Abdullahi, M. Characteristics of Wood ASH /OPC Concrete. Leonardo Electron Journal of practices and technology, 8 (2006), 9-16

[2] Adesanya, D.A., and Raheem A.A.. A study of the workability and compressive strength characteristics of cement composite sandcrete blocks containing sawdust ash. Construction and building materials, 23 (2009), 311-317 https://doi.org/10.1016/j.conbuildmat.2007.12.004

[3] Adewuyi, A.P., and Ola, S.A. Application of wood ash as partial replacement of cement in the production of sandcrete blocks. Journal of civil engineering and construction technology, 7 (2016) 9-16

[4] Anosike, M.N. and Oyebade A.A. "Sandcrete Blocks and Quality Management in Nigeria Building Industry", Journal of Engineering, Project and Production Management, 2 (2011) 37-46 https://doi.org/10.32738/JEPPM.201201.0005

[5] Anthony, B.S, Olabosipo I. F, Adewuyi, P. A, Musibau, A. A, sandcrete block and brick production in Nigeria - prospects and challenges" (2015)

[6] ASTM C618-12. Standard specification for coal fly ash and raw or calcined natural pozzolan for use in concrete. ASTM International. (2012)

[7] ASTM C 618. Standard specification for coal fly ash and raw or calcined natural pozzolan for use in concrete. American society for testing and materials (1994)

[8] British Standards Institution (BSI). BS 5628-1: code of practice for the use of unreinforced mansory. BSI Standard publication. (2005)

Advances in Cement and Concrete
Materials Research Proceedings 51 (2025) 39-48

Materials Research Forum LLC
https://doi.org/10.21741/9781644903537-5

[9] British Standards Institution. Specification for aggregates from natural sources for concrete. London, (1992)

[10] Cheah, C.B., and Ramli, M. The implementation of wood ash as a partial cement replacement material in the production of structural grade concrete and mortar: an overview. Resources, conservation and recycling,7 (2011), 669-685 https://doi.org/10.1016/j.resconrec.2011.02.002

[11] Leroy, M.N., Dupont, F. C., Rose, A., Joseph, N. and Bienenu, J.Density and strength of mortal made with the misture of woodash, crushed gneiss and river sand as fine aggregate. Journal of materials science and chemical engineering, 6 (2018), 109-120 https://doi.org/10.4236/msce.2018.64012

[12] National Association of Home Builders (NAHB). Soaring construction costs drop housing affordability to loweset level in a decade, (2022)

[13] NIS 087, Nigerian Industrial Standard: Standard for Sandcrete Blocks", Lagos, Nigeria, Standards Organisation of Nigeria (2000)

[14] NIS 87, Nigerian Industrial Standard Standard for sandcrete blocks. Nigerian Industrial Standard approved by The Standard Organisation of Nigeria (SON) (2004)

[15] NIS 444, Nigeria Industrial Standard. Code of Practice for Sanitary Installations: part 1-code of practice for sanitary installations in buildings. Lagos, Nigeria: standard organization of Nigeria" 2003

[16] NIS 554:, "Nigerian Standard for Drinking Water Quality," ed. 2007

[17] Oyekan, G. L. and O. M. Kamiyo. A study on the engineering properties of sandcrete blocks produced with rice husk ash blended cement. Journal of Engineering and Technology Research, 3 (2011), 88-98

[18] Popoola, O.C, Ayegbokiki, S.T. and Gambo M.D, Study of Compressive, Strength Characteristics Of Hollow Sandcrete Blocks Partially Replaced By Saw Dust Ash. IOSR Journal of Engineering (IOSRJEN), (2015),30-34

[19] Powrie, W. Soil mechanics, Concepts and applications (2018) https://doi.org/10.1201/9781315275284

[20] Ragh U.K., Sharath, V.T, Naveen, Y.., Bharathkumar, and yogesha, B.S, Experimental investigation on partial replacement of cement by mesquite (prosopis julifloria) wood ash in concrete international journal of scientific research and development,6 (2017),

[21] Raheem A.A., Adesanya,D.A., and Olofinnade, O.M, The effect of sawdust ash on the mechanical properties of sandcrete blocks. International journal of civil engineering and technology, 6 (2013), 196-204

[22] Raheem, A.A., Olumide, A. A, Wood ash from bread bakery as partial replacement for cement in concrete International Journal of Sustainable Construction Engineering & Technology,4 (2013) https://doi.org/10.5592/otmcj.2012.2.3

[23] Raheem, A. A. and Sulaiman, O. K, Saw Dust Ash as Partial Replacement for Cement in the Production of Sandcrete Hollow Blocks International Journal of Engineering Research and Applications (IJERA),4 (2013) 713-721

[24] Raheem, A. A, Comparism of the Quality of Sandcrete Blocks Produced by LAUTECH Block Industry with others within Ogbomoso Township, Science Focus, 11 (2006), 103-108.

[25] Raheem, A. A., Momoh, A. K. and Shoyingbe, A. A, Comparative Analysis of Sandcrete Hollow Blocks and Laterite Interlocking Blocks as Walling Elements, International Journal of Sustainable Construction Engineering & Technology, 3 (2012):79 - 88.

[26] Sebastian, A., Manapurath, A. S., Balachandran, D., Sebastian, D. M. and Philip, D, Partial replacement of cement with wood ash. International. Journal. Science. Technologies. Engineering. IJSTE, (2016), 666-670.

[27] Siddique,R, Utilization of wood ash in concrete manufacturing. Resources, conservation and Recycling, 67 (2012), 27-33. https://doi.org/10.1016/j.resconrec.2012.07.004

[28] Solomon I.A. and Jonas E. A, MILLET HUSK ASH AS PARTIAL REPLACEMENT OF CEMENT IN SANDCRETE BLOCK. International Research Journal of Engineering and Technology (IRJET) 04 (2017) 56-72

[29] Subramaniam, P., Subasinghe, K., Fonseka W.R.K, Wood ash as an effective raw material for concrete blocks Int. J.Res. Eng. Technology, 4 (2015), 1163-2319

[30] Svoboda, P. and Prochazka, M, Outdoor earthen plasters, Organisation, Technology and Management in Construction: an International Journal 4 (2012), pp.420-423. https://doi.org/10.5592/otmcj.2012.1.7

[31] Tahomah, A., Datok, E.P. and Jambol, D.D, Guinea Corn Husk Ash as Partial Replacement of Cement In Hollow Sandcrete Block Production. Journal of Emerging Trends in Engineering and Applied Sciences (JETEAS), 8 (2017),263-268

[32] Vilane, B.R.T., Vrbos, N. and Innocent, S.M, The effect of wood ash blending on the compressive strength of concrete blocks. Journals of agricultural science and engineering, 6 (2021), pp 43-47

Advances in Cement and Concrete
Materials Research Proceedings 51 (2025) 49-57

Materials Research Forum LLC
https://doi.org/10.21741/9781644903537-6

Effect of calcium nitrate on the workability and compressive strength of wood ash blended cement concrete

Akeem Ayinde RAHEEM[1,a], Oludayo Ajani AKINTOLA[2,b*],
Isaiah Oluwafemi OGUNTOLA[3,c] and Favour Wuraola KOLAWOLE[2,d]

[1]Department of Building, Ladoke Akintola University of Technology, Ogbomoso, Nigeria

[2]Department of Civil Engineering, Ladoke Akintola University of Technology, Ogbomoso, Nigeria

[3]Nigerian Building and Road Research Institute, Km10, Idiroko road, Ota, Ogun State, Nigeria

[a]aaraheem@lautech.edu.ng, [b]oludayo.akintola@gmail.com, [c]oguntolaisaiah@nbrri.gov.ng,
[d]kolawoleoludara@gmail.com

Keywords: Wood Ash, Calcium Nitrate, Supplementary Cementitious Materials, Workability, Compressive Strength

Abstract. Wood ash (WA) is emerging as a potential Supplementary Cementitious Material (SCM). Incorporation of WA as a partial replacement for cement in blended cement concrete presents environmental and economic benefits. This study examines the effect of calcium nitrate on the workability and compressive strength of blended cement concrete with a view to enhance its performance. The chemical composition of WA, CN and OPC were analyzed using XRF. Wood Ash was substituted at 10% while CN was replaced at 1,2,3,4 and 5% by weight of OPC. The slump and compacting factor of the fresh concrete were determined. The compressive strength was determined after 7, 14, 21, 28, 56 and 90 days of curing. Wood Ash was found to be a class C pozzolanic material as the sum of SiO_2, Al_2O_3 and Fe_2O_3 were 40.50%. The CaO content present in WA, CN and OPC were 46.27, 81.95 and 68.76%, respectively. The slump and compacting factor decrease from 39 to 19 mm and from 0.94 to 0.74 mm, respectively. The compressive strength of WA-CN varied from 14.08 to 46.24 N/mm^2. It was observed that inclusion of CN and WA improved the workability and compressive strength of the blended concrete with 4% CN being the optimum.

Introduction

In recent years, there has been a growing awareness of the environmental impact of pollution generated by cement manufacturing, prompting the need for sustainable alternatives to traditional cement production. Cement manufacturing is associated with significant amount of carbon dioxide (CO_2), all of which have adverse health effects, particularly on vulnerable populations such as infants, the elderly, and individuals with respiratory conditions like asthma, emphysema, or bronchitis.

Research has extensively explored the use of various pozzolans, such as wood ash, corn cob ash, sawdust ash, rice husk ash, coconut fiber ash, palm kernel shell ash and neem seed husk ash as replacements for cement rather than as additional components in concrete production [17,18,19,20,]. Pozzolanic cements are typically blends of ordinary Portland cement (OPC) with either natural or artificial pozzolans. Natural pozzolans are primarily of volcanic origin, though some diatomaceous earths also qualify, while artificial pozzolans include materials such as fly ash, burned clays, and shales [4]. Incorporating wood ash into concrete could therefore play a crucial role in advancing sustainable construction practices [7].

Blended cement, which is produced by incorporating Supplementary Cementitious Materials (SCMs) with Ordinary Portland Cement (OPC), offers a more sustainable alternative. SCMs such as Wood Ash, Rice Husk Ash, Fly Ash, Slag, Silica Fume, and Volcanic Ash are introduced during the grinding stage of cement production. The use of blended cement provides numerous advantages, including improved workability, durability, and strength, alongside enhanced construction outcomes. Moreover, its eco-friendly nature, characterized by lower carbon emissions during manufacturing, makes it an ideal choice for sustainable construction. The economic benefits also render blended cement a cost-effective option for building projects [8,10].

Calcium nitrate (CN) is commonly used as a raw material in the production of chloride-free accelerating admixtures for concretes containing Portland cement [24]. In the present study, the impact of (CN) on concretes made with blended Portland cements was examined, with comparisons drawn against concrete made with pure Portland cement. According to [13], the workability of a non-air-entrained concrete containing volcanic ash as a partial replacement for ordinary Portland cement, combined with calcium nitrate, exhibited slump values ranging from 63 mm to 75 mm, indicating satisfactory workability without segregation or excessive bleeding.

Materials and Methods
The materials used and the adopted method in carrying out this research were discussed in this section

Materials
Wood ash (WA) was sourced from Ladoke Akintola University of Technology bread bakery in Ogbomoso at a Latitude of $N6^040^I52^{II}$ and Longitude $E3^09^I35^{II}$. Ordinary Portland Cement (OPC), fine aggregate and coarse aggregate were sourced in Sango Ota, Ogun State. Calcium Nitrate(CN) was purchase from Videb Chemical Nigeria Limited, Farida Adeleke Market, Osogbo, Osun State. Portable water was obtained at the Building Research Laboratory of the Nigerian Building and Road Research Institute (NBRRI), Ota, Ogun State where the practical is performed.

Testing of materials
Characterization of WA, CN and OPC were examined using X-ray fluorescence analysis which was carried out at the Petroleum Reservoir Laboratory under the department of Chemical and Petroleum Engineering Afe Babalola University, Ado-Ekiti, Ekiti. The oxides composition such as silica (SiO_2), alumina (Al_2O_3), ferric oxide (Fe_2O_3), calcium oxide (CaO), magnesium oxide (MgO), Sulphur trioxide (SO_3) etc. were determined in line with [17,21] among several authors were obtained in this study.

The physical properties of WA, river sand (fine aggregate) and granite (coarse aggregate) determined were particle size distribution and specific gravity in line with the existing standard. These were carried out at the building laboratory complex, Nigerian Building and Road Research Institute (NBRRI), Idiroko road, Sango-Ota, Ogun State.

Concrete mix proportion
The concrete mix ratio of 1:2:4 and water-cement ratio of 0.6 was adopted in the study similar to Oriola et al.,2021. According to [16,22], the optimum WA content in blended concrete to enhance strength and workability is 10%. This study adopted a fixed WA content of 10%. Concrete mixing was carried out manually according to BS 1881-125 (2013). Table 1 shows the batching of material for production of concrete specimen. Control represent specimen without WA and CN while control 2 represent sample with only WA content. The remaining samples contained 10% WA content and varying percentage of CN from 1 to 5%.

Advances in Cement and Concrete
Materials Research Proceedings 51 (2025) 49-57

Materials Research Forum LLC
https://doi.org/10.21741/9781644903537-6

Table 1. Batching of material for production of concrete specimen

Samples Identification	Specimen	OPC (kg)	WA (kg)	CN (kg)	FA (kg)	CA (kg)
A	100% OPC (Control 1)	50	0	0	100	200
B	90% OPC + 10% WA (Control 2)	45	5	0	100	200
C	89% OPC + 10% WA + 1%CN	44.5	5	0.5	100	200
D	88% OPC + 10% WA + 2%CN	44	5	1.0	100	200
E	87% OPC + 10% WA + 3%CN	43.5	5	1.5	100	200
F	86% OPC + 10% WA + 4%CN	43	5	2.0	100	200
G	85% OPC + 10% WA + 5%CN	42.5	5	2.5	100	200

FA= Fine Aggregate   CA=Coarse Aggregate

Specimen Preparation
The compressive strength concrete mixes were cast into cubes of 100 mm x 100 mm x 100 mm and were determined at 7, 14, 21, 28, 56 and 90 curing days. Three cubes were casted for each specimen so as to determine the average compressive strength. A total of 126 cubes were cast for Compressive strength test. This was performed in accordance with BS EN 12390-3:2019 standard.

Results and Discussion
The results and discussion on the research were discussed in this section.

Characterization of materials
The characterization of Wood Ash, Calcium Nitrate and Ordinary Portland Cement were analyzed using XRF analysis and the results are shown in Table 2. The chemical analysis denotes that Wood Ash belong to Class C pozzolan because the sum of SiO_2, Al_2O_3 and Fe_2O_3 ≤ 50% with LOI < 6% as specified by ASTM C 618 (2019) which states that class C pozzolan have higher calcium content and lower silica and alumina content. Also, XRF analysis of CaO present in Calcium Nitrate and OPC as presented in Table 2 were 81.95 and 68.76%, respectively. Wood Ash satisfies the maximum value of 5% specified by ASTM C618-12 (2012) for the SO_3 of cementitious material with a value of 0.5%.

Table 2:   Chemical Composition of WA, CN and OPC

Chemical Constituent	Percentage Composition (%)		
	Wood Ash	Calcium Nitrate	OPC
SiO_2	39.71	2.05	18.65
Al_2O_3	0.59	0.60	3.24
Fe_2O_3	0.20	5.01	3.24
CaO	46.27	81.95	68.76
MgO	1.70	5.61	1.82
Na_2O_3	0.41	0.43	0.37
K_2O	3.02	1.10	1.65
P_2O_5	2.13	0.05	-
SO_3	0.50	1.60	2.90
LOI	4.34	1.49	1.73

$SiO_2 + Al_2O_3 + Fe_2O_3$ = 40.50 < 50%

Physical properties of WA, sand and coarse aggregate

This section outlines the results of the laboratory investigation into the physical properties of WA, sand, and aggregate, covering aspects such as particle size distribution and specific gravity.

Particle size distribution

The results of the particle size distribution (PSD) analysis carried out in this research work are graphically presented in Figure 1. The grading curve of aggregates shows the coefficient of uniformity (Cu) and coefficient of curvature (Cc) values to be 3.4 and 1.6 for fine aggregate, 2.0 and 1.09 for coarse aggregate. The Cu value for fine aggregate is less than 4 and Cc value is between 1 to 3 as specified by unified soil classification system (USCS) as postulated in ASTM D2487 (2011). Thus, it can be categorized as well graded sand and gravelly sands. The Cu gives an indication of the spread of particle size value [15]. The Cu for coarse aggregate is 1.2. This can be classified as uniformly graded. Wood ash (WA) used was sieve through sieve 425 μm. Based on these assessments, WA, fine aggregate and coarse aggregate are suitable ingredients for making quality concrete [5].

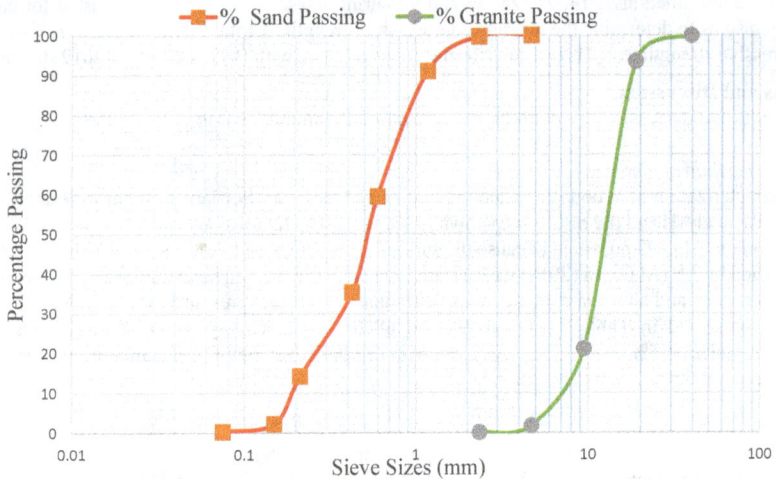

Fig. 1. Particle size analysis of sand and granite

Specific gravity

The result of specific gravity of the materials is as presented in Table 3. The specific gravity of WA is lower than that of river sand and granite. Therefore, WA is expected to produce a less dense concrete with better thermal, insulative properties and significant reduction in dead weight than the conventional aggregate when use in concrete [23]. The specific gravity value of WA (1.97 g/cm3) falls outside the range of 2.5-3.0g/cm3 specified by BS EN ISO 10545 (1997) for normal weight aggregate and can be regarded as low weight aggregate. The specific gravity of WA (1.97 g/cm^3) is in correlation with the result of [9]. The specific gravity values obtained for granite (2.71 g/cm^3) and river sand (2.62 g/cm^3) falls within the range of 2.5-3.0 g/cm3 as specified by [12]. It also in accordance with ASTM C33/C33M (2023) standard specification for concrete aggregates

Advances in Cement and Concrete
Materials Research Proceedings 51 (2025) 49-57

Materials Research Forum LLC
https://doi.org/10.21741/9781644903537-6

which acknowledges that normal-weight of aggregates commonly have specific gravities between 2.4 g/cm³ and 2.9 g/cm³.

Table 3:   Specific gravity of WA, OPC and aggregates

Materials		Specific Gravity
WA		1.97
OPC		3.10
Fine Aggregate (Sand)	2.62	
Coarse Aggregate (Granite)		2.71

Workability
This section discusses the workability (slump and compacting factor) of the control samples and varying WA-CN blended cement.

Slump
The result of the slump test is presented in Figure 2. The result indicates that the inclusion of WA-CN as a substitute material reduces collapse. Sample A and B specimen had slump value of 39 mm and 37 mm respectively while the varying WA-CN mixes at interval 1 o 5% CN were 36, 34, 30, 24 and 19mm respectively. This shows that the slump values decrease from 39 to 37mm after the addition of 10% WA at B while later reduces from 36 to 19mm as the percentage of CN content increases from 1 to 5%. This is in accordance with ASTM C143 and BS1882 part 2. It was noticed that as the percentage of calcium nitrate content increases, the slump values reduces. Therefore, in other to attain the required workability, mixes containing WA-CN will require more water than conventional mixtures.

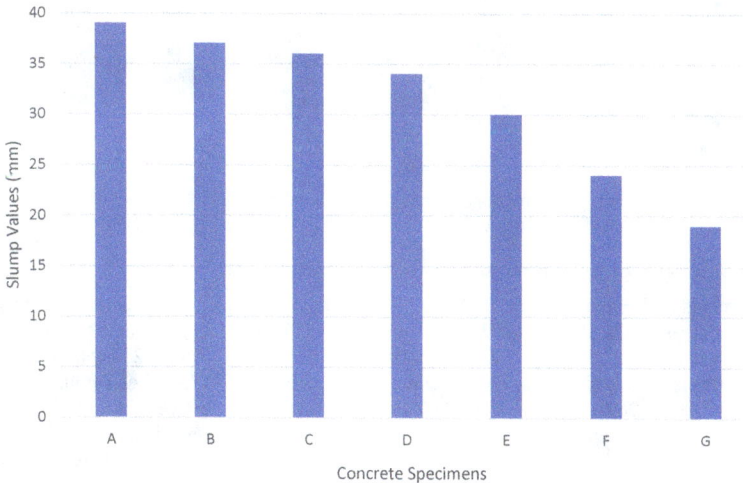

Figure 2: Slump of WA-CN Concrete

Compacting factor

The result for compacting factor test as obtained in Figure 3 shows that control specimen A has a value of 0.94. The specimens with varying CN content from 1 to 5 % have values of 0.91, 0.87, 0.83, 0.78 and 0.74, respectively. This shows that compacting factor decreases as the CN content increases. This is supported with the result of [11] that categorized compacting factor values as very low (0.78), low (0.85), medium (0.92) and high (0.95) in accordance with building research establishment. This result shows that sample A could be categorized in high degree workability. Sample B categories as medium degree while sample C, and D falls in low degree of workability and sample E and F fall in very low degree of workability. As the addition of CN increases the concrete workability becomes low.

Compressive Strength of WA-CN blended cement concrete

The results of Compressive Strength of WA-CN blended cement concrete tested at 7, 14, 21, 28, 56 and 90 days were presented in Figure 4. Calcium nitrate was added at 1 to 5% dosage with 10% WA content. The compressive strength at 7 days shows that strength of sample A

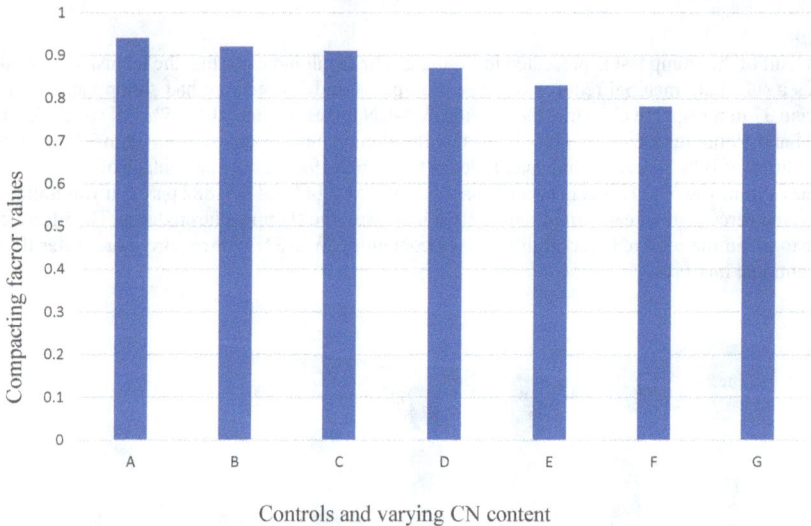

Controls and varying CN content

Figure 3: Compacting Factor of WA-CN Concrete

(19.47 N/mm^2) is higher than sample B (16.32 N/mm^2), then starts increasing as the percentage CN contents increases. The strength of WA-CN ranged from 14.08 to 25.12 N/mm^2 as CN content varies from 1 to 5%, respectively. Similar trends were observed at 14 and 21days where the strength of sample A is higher than B. The compressive strength of WA-CN at 14 and 21days varied from 16.06 to 26.01 N/mm^2 and 19.07 to 28.30 N/mm^2, respectively. At 28 days, the compressive strength for sample A and B are 26.45 and 21.54 N/mm^2, respectively while the WA-CN ranged from 22.38 to 29.75 N/mm^2. At 56 days, compressive strength for sample A and B are 29.3 and 24.78 N/mm^2, respectively while the WA-CN ranged from 30.61 to 37.01 N/mm^2 as the CN contents varies from 1 to 5% respectively. At 90days the result of the compressive strength for sample A and B were 38.12 and 31.06 N/mm^2 respectively. The compressive strength of WA-CN at 90days varied from 37.45 to 46.24 N/mm^2 as the CN contents varies from 1 to 5%, respectively. The inclusion of CN in the mixture was noted to enhance the strength performance of the concrete. At early age, the compressive strength of WA-CN increases with curing age. The strength increases with increase in the quantity of calcium nitrate from 4% upward. Therefore, a peak compressive strength was attained at 5% dosage of CN. This indicate that 5% dosages of calcium nitrate inclusion in the mixture is considered the peak for improving the compressive strength of concrete incorporating with 10% WA content. The result is in accordance with BS EN 12390-3:2019 and also in correlation with the result of [13]

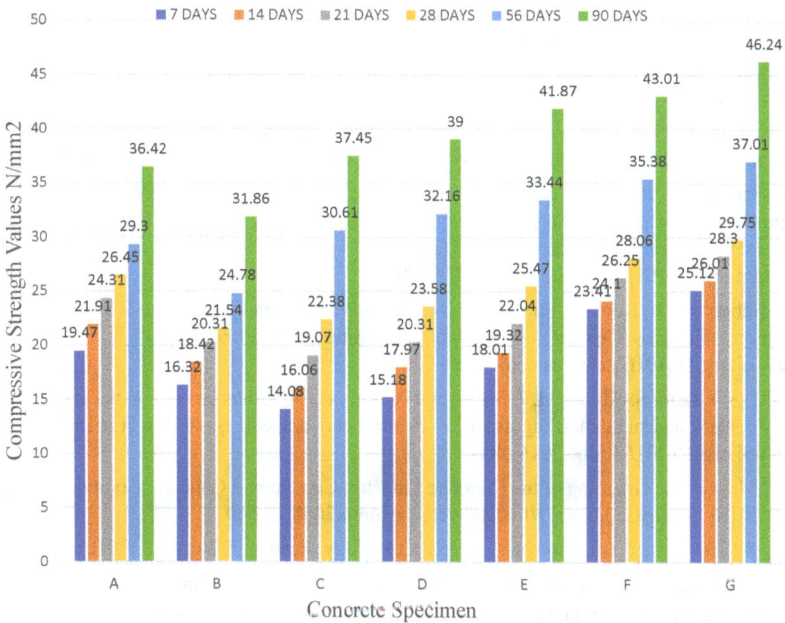

Figure 4:   *Compressive Strength of WA- CN blended cement concrete.*

Conclusions
From the findings in this study, the following conclusions are made

- Wood ash belongs to class C pozzolan as the sum of silica, alumina and ferric oxide is less than 50% while Calcium nitrate was found to be a notable admixture with present of high CaO content;

- The workability of fresh concrete Wood ash decreases with increase in calcium nitrate content;

- Calcium nitrate addition improved the compressive strength properties of WA-CN concrete.

References

[1] ASTM C33/C33M. Standard specification for concrete aggregates. ASTM International; (2023).

[2] ASTM C618-12. Standard specification for coal fly ash and raw or calcined natural pozzolan for use in concrete. ASTM International, West Conshohocken, PA; (2012). DOI: 10.1520/C0618-12. https://doi.org/10.1520/C0618-12

[3] ASTM D2487. Standard practice for classification of soils for engineering purposes. ASTM International, West Conshohocken, PA. (2011).

[4] M. Blaise, R. Jones and K. Smith. Advances in the use of Natural and Artificial Pozzolans in Cementitious Materials. Cement and Concrete Research, 120 (2023) 95-107.

[5] J. E. Bowles. Engineering properties of soils and their measurements. 4th ed. New York: McGraw-Hill.(1992).

[6] BS EN ISO 10545. Ceramic tiles. Determination of water absorption, apparent porosity, apparent relative density, and bulk-density. British Standards Institution, London. (1997).

[7] K. Ganesan, K. Rajagopal and K. Thangavel. Rice Husk Ash Blended Cement: Assessment of Optimal Level of Replacement for Strength and Permeability Properties of Concrete. Construction and Building Materials, 22,8 (2008) 1675-1683. https://doi.org/10.1016/j.conbuildmat.2007.06.011

[8] G. Habert and N. Roussel. Study of two Concrete Mix-design Strategies to Reach Carbon Mitigation Objectives. Cement and Concrete Composites, 31,6 (2009) 397-402. https://doi.org/10.1016/j.cemconcomp.2009.04.001

[9] S. M. S. Kazmi, S. Abbas, M. J. Munir and A. Khitab. Exploratory study on the effect of waste rice husk and sugarcane bagasse ashes in burnt clay bricks. J Build Eng 7(2016) 372-378. https://doi.org/10.1016/j.jobe.2016.08.001

[10] C. Meyer. The Greening of the Concrete Industry. Cement and Concrete Composites, 31,8 (2009) 601-605. https://doi.org/10.1016/j.cemconcomp.2008.12.010

[11] A. M. Neville and J. J. Brooks. Concrete technology. Pearson Education. (2010).

[12] A. M. Neville. Properties of concrete. 5th ed. Pearson Education Limited. (2011).

[13] E. B. Ogunbode and I. O. Hassan. Effect of addition of calcium nitrate on selected properties of concrete containing volcanic ash. Leonardo Electron J Pract Technol. 19 (2011) 29-38.

[14] K. O. Oriola, A. A. Raheem, M. A. Kareem and R. Abdulwahab. Assessment of workability and compressive strength of rice husk ash-blended palm kernel shell. J Civil Environ Stud. 7,1 (2021). https://doi.org/10.36108/laujoces/1202.70.0101

[15] A. A. Raheem and B. D. Ikotun. Incorporation of agricultural residues as partial substitution for cement in concrete and mortar: A review. J Build Eng 31 (2020). https://doi.org/10.1016/j.jobe.2020.101428

[16] A. A. Raheem and M. A Kareem. Optimal raw material mix for the production of rice husk ash blended cement. Int J Sustain Constr Eng Technol 2017a;7(2).

[17] A. A. Raheem and M. A. Kareem. Chemical composition and physical characteristics of rice husk ash blended cement. Int J Eng Res Afr 2017b;32:25-35. https://doi.org/10.4028/www.scientific.net/JERA.32.25

[18] A. A. Raheem, O. T. Adenuga and A. Shittu. Effect of Sawdust Ash on the Properties of Concrete. Journal of Sustainable Development, 5,4 (2012) 84-90.

[19] A. A. Raheem and O. T. Adenuga. Wood Ash as an Additive in Concrete. Journal of Construction Engineering and Management, 139,2 (2013) 179-184.

[20] A. A. Raheem and D. A. Adesanya. Production of Ash from Corn Cob and its Effect on the Compressive Strength of Concrete. Journal of Environmental Design and Management, 2,1 (2009) 45-50.

[21] A. Z. Seyed, A. Farshad, D. Farzan and A. Mojtaba. Rice husk ash as a partial replacement of cement in high strength concrete containing micro silica. Case Study Constr Mater. 7 (2017) 73-81. https://doi.org/10.1016/j.cscm.2017.05.001

[22] R. Swetha, M. S. Rani and U. Raju. Experimental works on self-compacting concrete by partial replacement of rice husk ash with subjected to acid attack. Int J Innov Technol Explor Eng 9 (2020) 664-667. https://doi.org/10.35940/ijitee.C8334.019320

[23] D. C. L. Teo, M. A. Mannan, V. J. Kurian and C. Ganapathy. Lightweight concrete made from oil palm shell (OPS): Structural bond and durability properties. Build Environ. 42 (2007) 2614-2621. https://doi.org/10.1016/j.buildenv.2006.06.013

[24] J. Wolfram, A. M. Smith and R. J. Lee. The Effect of Calcium Nitrate on Concrete Hydration and Strength Development. Journal of Building Materials, 29,3 (2015) 54-67.

Advances in Cement and Concrete
Materials Research Proceedings 51 (2025) 58-67

Materials Research Forum LLC
https://doi.org/10.21741/9781644903537-7

Assessment of the effect of different sand grading on strength properties of metakaolin blended cement mortar

Solomon Olalere AJAMU[1,a], Rasheed ABDULWAHAB[2,b*],
Emmanuel Olatunde IBIWOYE[3,c], Toheed Ayomide OYEBAMIJI[1,d],
Ayomide Faith USMANF[1, e], and Akinbobola Joseph AKINDEJOYE[1, f]

[1]Department of Civil Engineering, Ladoke Akintola University of Technology, Ogbomoso Oyo State Nigeria

[2]Department of Civil Engineering, Kwara State University, Malete Kwara State Nigeria

[3]Kwara State Polytechnic, Ilorin Kwara State Nigeria

[a]soajamu@lautech.edu.ng, [b]abdulwahab.rasheed@kwasu.edu.ng,
[c]ibiwoye.e@kwarastatepolytechnic.edu.ng, [d]toyebamiji2018@gmail.com,
[e]ayomideu305@gmail.com, [f]akinbobola2002@gmail.com

Keywords: Metakaolin, Sand Grading, Compressive Strength, Flexural Strength, Cement Mortar, Sustainability

Abstract. Aggregate size has effects on the properties of cement composite. However, determination of sand grading is not usually put in to consideration in producing cement mortars for different application. Effects of different sand grading on the strength properties of metakaolin blended cement mortar have also not been well evaluated. In this study, effect of sand grading on the mechanical properties of metakaolin blended cement mortar, towards ensuring the development of durable and sustainable cement composite was investigated. Kaolin was sourced from Oyo State, South-West Nigeria. It was calcinated at 700°C to form metakaolin (MK) and characterized using X-Ray Fluorescence to determine its oxide composition. A blended cement produced with 15 to 25% replacement of Portland Limestone Cement (PLC) with the MK was used to produce mortar with different sand grading of fine, medium and coarse. Workability tests were carried out on freshly prepared mortar specimen with the three different sand grading. Water absorption, compressive strength and flexural strength tests on the different mortar specimen were carried out at 7, 14, 21, and 28 days, respectively. The addition of SiO_2, Al_2O_3 and Fe_2O_3 of MK was found to be 79.23%. The grading variation are; Fine Sand (0.053 mm to 0.212 mm), Medium Sand (0.212 mm to 0.600 mm) and Coarse Sand (0.600 mm and 2.00). Workability of the mortar decreases with increasing particle sizes of sand and percentage replacement of PLC with MK. Fine sand absorbed more water with the highest at 25% MK replacement. Mortars specimen produced with Medium sand gave the highest compressive strength (15.11, 14.89 and 15.56 N/mm^2) and flexural strength (5.616, 6.181 and 5.828 N/mm^2) respectively for each of the 15, 20 and 25% replacement with MK. This study recommends the application of mortar containing medium sized sand with 20% MK for construction purposes.

Introduction

Cement mortar, a fundamental material used in construction is a very important component of concrete, and its properties impact the overall performance of the concrete structure significantly. Grading of sand is one of the key factors influencing the properties of cement mortar and it is a measure of the distribution of particle sizes within the sand, which can affect the workability, strength, and durability of the mortar. A sand with a balanced proportion of fine and coarse particles (well-graded sand) can improve the workability of the mortar, making it easier to mix and apply [1]. Additionally, a well-graded sand can enhance the strength of the mortar by reducing

Advances in Cement and Concrete
Materials Research Proceedings 51 (2025) 58-67

Materials Research Forum LLC
https://doi.org/10.21741/9781644903537-7

the water-cement ratio and improving the packing density of the particles [2]. Sand grading has influence on the performance of the mortar because a well-graded sand can reduce the permeability of the mortar, making it less susceptible to degradation from water and aggressive substances [3]. Furthermore, the resistance of the mortar to freeze-thaw cycles, corrosion, and chemical attack can be improved by a well-graded sand, [4, 5, 6]. It is possible to achieve high-performance mortar that meets the demands of various applications by selecting a well-graded sand for the production of the mortar and by optimizing the mortar mix design.

Calcining of kaolin clay at high temperatures in the range of 600-800°C gives a supplementary cementitious material (SCM) called Metakaolin [7]. Calcination transforms the kaolin into a highly reactive, amorphous aluminosilicate material. Metakaolin is highly reactive, making it an effective pozzolan for improving the strength and durability of concrete [8]. It improves the workability of concrete by reducing the water demand and improving the flowability [9]. It also enhances the durability of concrete by reducing the permeability, improves the resistance to chemical attack, and reduces the risk of alkali-silica reaction. A type of blended cement that combines ordinary Portland cement (OPC) with metakaolin is referred to as Metakaolin blended cement. The addition of MK can improve the performance of the cement in several ways such as improved strength, durability and sustainability [7]. In this study, effect of MK blended cement on the properties of mortar produced with different sand gradings were investigated

Literature Review
Sand grading and its impact on cement mortar
Sand grading plays a crucial role in determining the properties of cement mortar. The grading of sand refers to the distribution of particle sizes within the sand, which can affect the workability, strength, and durability of the mortar. A well-graded sand (balanced proportion of fine and coarse particles) can improve the workability of the mortar, making it easier to mix and apply [1]. On the other hand, a poorly graded sand (either too fine or too coarse) can lead to reduced workability and increased water demand [2]. The grading of sand also affects the strength of the mortar. A well-graded sand can produce a stronger mortar, as it allows for better packing and bonding between the cement particles [3]. A poorly graded sand, on the other hand, can lead to reduced strength and increased porosity [4]. The durability of the mortar is also influenced by the grading of sand. A well-graded sand can improve the resistance of the mortar to freeze-thaw cycles, chemical attack, and abrasion [6]. A poorly graded sand, on the other hand, can lead to reduced durability and increased risk of degradation [5]. Sand grading plays a crucial role in determining the properties and performance of cement mortar [1]. This significantly affects the packing density, workability, and ultimately the strength of the mortar [2]. Understanding the impact of sand grading on these factors is essential for optimizing the mix design and achieving desired outcomes in construction projects.

Workability describes the ease with which mortar can be mixed, placed, and finished. Sand grading has a direct impact on the workability of the mix [10]. Well-graded sand improves workability by providing a smooth, cohesive mix that is easy to handle and apply. Poorly graded sand, particularly if it is too fine or too coarse, can lead to a harsh mix that is difficult to work with and may segregate.

The strength of mortar is largely influenced by the sand grading. Well-graded sand contributes to a stronger mortar by enhancing the packing density and reducing the void content [3]. The presence of a wide range of particle sizes ensures that the sand particles interlock effectively, providing a solid and stable structure. This interlocking mechanism enhances the overall strength and load-bearing capacity of the mortar. Uniformly graded sand consists predominantly of particles of similar size. This type of sand tends to have lower packing density because there are significant voids between the particles [4].

Advances in Cement and Concrete
Materials Research Proceedings 51 (2025) 58-67

Materials Research Forum LLC
https://doi.org/10.21741/9781644903537-7

To achieve optimal properties in cement mortar, it's essential to select a well-graded sand that meets the requirements of the specific application. The ideal sand grading will depend on factors such as the type of cement, aggregate, and intended use of the mortar [10].

Materials and Method

Material
The materials that were used for the study are metakaolin, sand, Dangote brand of Portland Limestone Cement (PLC), following materials were used in the experimental design.

Metakaolin
High-reactivity kaolin (Plate 1a) from fused fatusi, Ibadan 110115, Oyo State, Nigeria (Lat N7°18'28.04076", E3°55'26.13252" and alt 207yd a.s.l) was used for the research work conforming to [11]. The collected kaolin with initial moisture content of 4.27% was air dried for 24 hours (Plate 1b) and was taken to Geology Department, University of Ilorin for calcination (at 700°C for 2 hours). The Metakaolin was ground to fine powdery form and those passing 75μm sieve was used. The metakaolin was taken to Afe Babalola University, Ekiti State, Nigeria, for chemical analysis.

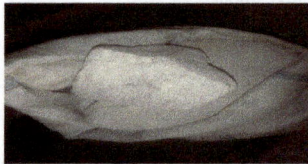

Plate 1a: Sample of Raw Kaolin   Plate 1b: Air drying of Kaolin

Sand
Sharp sand obtained from LAUTECH area, Ogbomoso was used as the fine aggregate. The sand was carefully collected washed and wet sieved to separate the sand from any unwanted materials and impurities and then sun dried. The sand was sieved through set of sieves with the use of sieve shaker to obtain the particle size distribution of the washed sand and thereafter graded. The grading variation are: Fine Sand (0.053 to 0.212 mm), medium sand (0.212 to 0.600 mm) and coarse sand (0.600 to 2.000 mm), The samples of the different sand gradings are as presented in Plate 2.

Water
Clean water free from impurities was used for the research work conforming to [12]. It was used for mixing the mortar and for curing the samples.

Cement
Ordinary Portland cement manufactured by Dangote Cement Company, conforming to [13]. The Ordinary Portland cement (Dangote Brand) used was obtained from a local retailer in Ogbomoso.

Fine sand   Medium sand   Coarse sand

Plate 2: Standardized sand

Advances in Cement and Concrete
Materials Research Proceedings 51 (2025) 58-67

Materials Research Forum LLC
https://doi.org/10.21741/9781644903537-7

Method

Metakaolin was grinded and sieved accordingly. MK passing 75 µm sieve was used for the production of blended cement. In the production of mortar, Metakaolin was used to replace cement in varying percentages of 15, 20, and 25 %, respectively. The mix ratio used was 1:3 (binder and sand) with water to binder ratio of 0.45 and mortar with no MK served as the control. The details of the mortar mixtures used in the tests are as presented in Table 1. Compressive strength and flexural strength tests were carried out on 150 × 150 × 150 mm mortar cubes and 40 × 40 × 160 mm mortar prisms, respectively. These tests were carried out in accordance with the procedures outlined in [14] and [15] respectively. The compressive strength and flexural strength were respectively calculated using equation (1) and equation (2). The specimens were cast in three layers; each layer being tamped with 35 strokes of the tamping rod spread uniformly over the cross section of the mould. The top of each mould was smoothened and leveled. The moulds and their contents were kept in the laboratory at temperature of 23°C ± 2°C and average relative humidity of 90% for 24 hours. All specimens were demould after 24 hours and cured in water at 23°C ± 2°C. The Compressive strength and flexural strength were determined at curing ages 7, 14, 21 and 28 days (Plate 3); using compression machine with maximum capacity 2000 kN and Flexural machine (Plate 4 a and b). The strength value was the average of three specimens.

Table 1*: Summary of the Mix Ratio*

Mix	Percentage Replacement (%)	Cement (kg)	Metakaolin (kg) (A)	Fine Sand (kg) (B)	Medium Sand (kg) (Sample 2)	Coarse Sand (kg) (Sample 3)	Water (kg)
C 25	0	15.11	0	45.1	45.1	45.1	6.9
1:3	15	12.84	2.27	45.1	45.1	45.1	6.9
W/C 0.45	20	12.08	3.02	45.1	45.1	45.1	6.9
	25	11.33	3.78	45.1	45.1	45.1	6.9

A = Control samples, B = Samples 1, C = Samples 2 and C = Samples 3

$$Cube\ strength = \frac{F}{A} \qquad (1)$$

Where F is the crushing load (N) and A is the Load Bearing area of the cube (mm²)

$$Flexural\ strength, R = \frac{3PL}{2bd^2} \qquad (2)$$

Where P is the load at failure (N), L is the beam span (mm), b the beam width (mm) and d is the depth (mm)

Plate 3: Curing of Mortar Beams and cubes in a curing tank

Advances in Cement and Concrete
Materials Research Proceedings 51 (2025) 58-67

Materials Research Forum LLC
https://doi.org/10.21741/9781644903537-7

Plate 4a: Compressive Strength Test

Plate 4b: Flexural Strength Test

Slump, compacting factor and density tests were carried out to check the effect of Metakaolin on the workability of fresh mortar with different sand grading (Plate 5). The tests were carried out in accordance with corresponding standards BS 1881-102 [16] for slump the test, and BS 1881-103 [17] for the compacting factor test. The rate of water absorption was determined for mortar at 28 days curing age.

Plate 5a: Slump test

Plate 5b: Compaction factor test

Results and Discusions

Chemical Composition

The elemental oxides composition present in the metakaolin are shown in Table 2. The results indicated that the percentage of SiO_2 present in the metakaolin is greater than the minimum requirement of natural pozzolan (minimum of 35%). The combined percentage of SiO_2, Al_2O_3 and Fe_2O_3 is more than 70%. This indicates that metakaolin is a good pozzolanic material in accordance with the requirements in [18]. Metakaolin produced from kaolin obtained from Fatusi, Ibadan falls under the category of **Class N** for Natural pozzolan since the sum of (SiO_2 + Al_2O_3 + Fe_2O_3) is greater than 70%. Loss of Ignition (2.885) did not exceed 10%. This indicates that the amount of unburned carbon and moisture content in the material is adequate according to the requirements in [18].

Table 2: Chemical Composition of Metakaolin

S/N	Chemical Constituents	Average (%)	S/N	Chemical Constituents	Average (%)
1	SiO_2	51.73	7	Na_2O	1.57
2	Al_2O3	16.45	8	P_2O_5	0.15
3	MgO	6.48	9	TiO_2	0.51
4	CaO	15.37	10	MnO	0.105
5	Fe_2O_3	5.57	11	NiO	0.01
6	K_2O	3.70	12	ZnO	0.01
7	Na_2O	1.57	13	CuO	0.01
8	P_2O_5	0.15	14	SO_3	0.125
9	TiO_2	0.51	15	LOI	2.885
				$\sum (SiO_2, Al_2O_3, Fe_2O_3)$	73.75

Advances in Cement and Concrete
Materials Research Proceedings 51 (2025) 58-67

Materials Research Forum LLC
https://doi.org/10.21741/9781644903537-7

Sieve Analysis of the washed sand

The results of the sieve analysis for washed sand used is as presented in Figure 1a. The coefficient of uniformity (Cu) and coefficient of gradation (Cc) for the sand was found to be 4.25 and 1.47 respectively. A Uniformity coefficient (C_u) of 4.25 indicates a relatively well-graded sand, as it is greater than the standard threshold of 3. It is also noticeable in the particle distribution curve, as the grain diameters of particles range uniformly across the sieve sizes. The curvature coefficient (C_c) of 1.47 falls within the acceptable range (1 to 3). This suggests a satisfactory distribution of particle sizes. Thus, the sand is classified as being well graded. Therefore, the fine, medium and coarse sand are suitable for making good mortar. Further sub-division of the sand gave fine: Fine Sand (0.053 to 0.212 mm), medium sand (0.212 to 0.600 mm) and coarse sand (0.600 to 2.000 mm) each of which was used to produce mortar specimen with MK blended cement containing different percentage of MK.

Workability

The results of the slump and compacting factor indicating the workability of the metakaolin mortar are shown in Figure 2 which shows that the slump increases with increase in metakaolin replacement for fine, medium and coarse sand respectively. The slump result falls under true slump with minimum change in the mortar height (within 15 – 35 mm) which is adequate for use.

The compacting factor also follows a similar trend. These results indicate that mortar containing Metakaolin beyond 15% replacement level becomes more workable as the metakaolin content increases meaning that less water is required to make the mortar workable. The lower water demand can be attributed to the finer metakaolin particles with filling role in metakaolin mix.

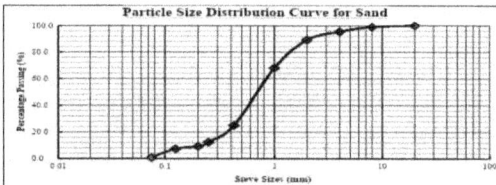

Figure 1a: Particle size distribution curve for the raw sand.

Water Absorption

Cube and beam water absorption for metakaolin blended cement mortar at 28days curing time is as presented in Figure 3a and b. Medium sand has the lowest rate of water absorption, and fine sand has the highest rate of water absorption, while coarse sand falls between the two. The rate of water absorption increased with metakaolin replacement for medium sand due to metakaolin's pozzolanic reaction which consumes calcium hydroxide, reducing pore volume and water absorption, whereas the rate of water absorption decreased with metakaolin replacement for fine and coarse sand. The highest rate of water absorption for fine, medium and coarse sand are; 11.58, 4.05 and 10.32 %, respectively for cubes and 4.23, 3.87and 4.0 %, respectively for beam.

Fine sand has higher rate of water absorption due to; larger surface area-to-volume ratio thereby increasing contact with water, smaller particles create more pores thereby allowing more water absorption, and higher porosity enables more water to penetrate. Medium sand has the lowest rate of water absorption due to; medium sand's more uniform grading reduces pores and water absorption, medium sand's lower porosity limits water penetration, medium sand's particles may have smoother surfaces, reducing water absorption, lower void ratio reduces water-filled spaces, and higher density decreases water absorption.

Advances in Cement and Concrete
Materials Research Proceedings 51 (2025) 58-67

Materials Research Forum LLC
https://doi.org/10.21741/9781644903537-7

According to ASTM C128, the standard rate of water absorption for fine, medium and coarse sand are; 5-10, 3-6 and 1-3 respectively. Exceptions and Variations in the result obtained when compared to ASTM C128 was variations in sand source and location. It was concluded that the main factors Influencing Water Absorption are; Particle size distribution, Particle shape and texture, Porosity and void ratio, Specific gravity and Grading and classification.

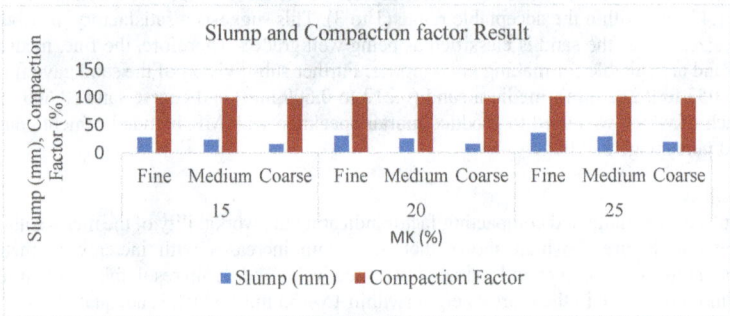

Figure 2: Slump and Compaction Factor result

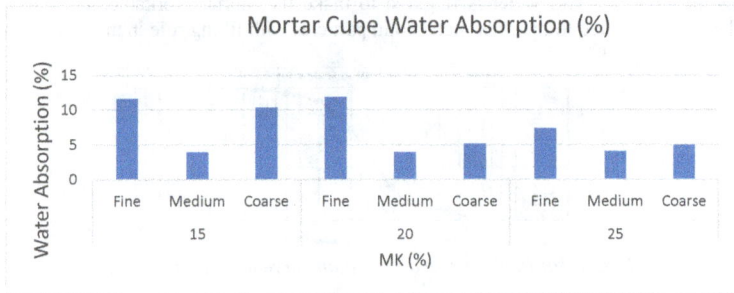

Figure 3a: Mortar Cube Water Absorption for MK blended cement mortar at 28 days

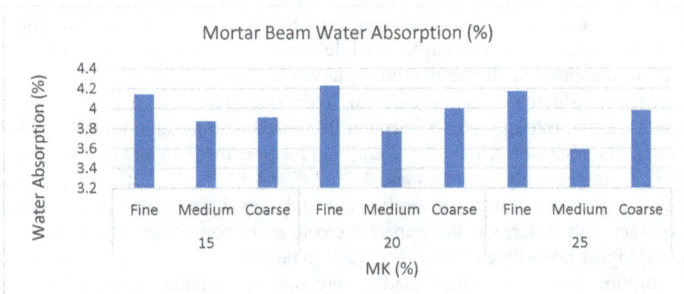

Figure 3b: Mortar Beam Water Absorption for metakaolin blended cement mortar at 28 days

Advances in Cement and Concrete
Materials Research Proceedings 51 (2025) 58-67

Materials Research Forum LLC
https://doi.org/10.21741/9781644903537-7

Compressive Strength

Compressive Strength of metakaolin blended cement mortar over curing days as shown in Table 9a and b and the graph as presented in Figure 4. The compressive strength of the metakaolin blended cement mortar increased with curing days and metakaolin replacement for medium and coarse sand, while the compressive strength of the metakaolin blended cement mortar increased with curing days and decreased with metakaolin replacement for fine sand. Highest compressive strength obtained for fine, medium and coarse sand are; 14.22, 15.56 and 14.67 N/mm², respectively. Medium sand has the highest compressive strength with respect to curing days and metakaolin replacement out of the three sand grades.

The compressive strength results of MK blended cement mortar in this study indicate that 20% metakaolin replacement provides the best performance across all sand grading. At 28 days, medium sand achieves the highest compressive strength of 15.56 N/mm², while fine sand exhibits the lowest strength (12.44 N/mm²), likely due to increased water reqirement, leading to higher porosity. Coarse sand shows moderate performance, with strengths around 14.67 N/mm² for the same MK replacement level.

These findings align with the ASTM C 618 standard, which requires pozzolanic materials to reach at least 75% of control strength. It was reported by [20] that 20% replacement yields 15-20% higher strength than standard cement mixes. In this study, the compressive strength of medium sand (15.56 N/mm²) for 20% replacement after 28 days matches these findings. Sabir et al. (2001): Showed that higher metakaolin content enhances the pozzolanic reaction but can limit strength gains beyond 20% replacement, which is also reflected in this results.

The variations in strength can be attributed to *Abrams' law*, which states that strength decreases as the water-cement ratio increases, and packing density theory, where medium sand's optimal particle distribution minimizes voids, resulting in denser and stronger mortar. In contrast, fine sand's high surface area increases water requirement, reducing strength due to excess porosity. Overall, the study confirms that 20% metakaolin replacement, particularly with medium sand, is ideal for achieving high compressive strength in cement mortar applications.

Flexural Strength

Flexural Strength of MK blended cement mortar at 28 day of curing for different sand grading used at varying percentage MK substitution is as shown in Figure 5. The highest flexural strength at 20 % MK for fine, medium and coarse sand are; 5.583, 6.181, and 5.854 N/mm², respectively. The highest flexural strength (6.181 N/mm^2) obtained for medium sand.

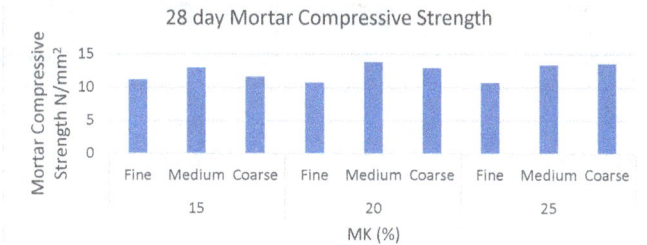

Figure 4: Compressive Strength of MK Blended Cement Mortar with varying Sand Grading and MK

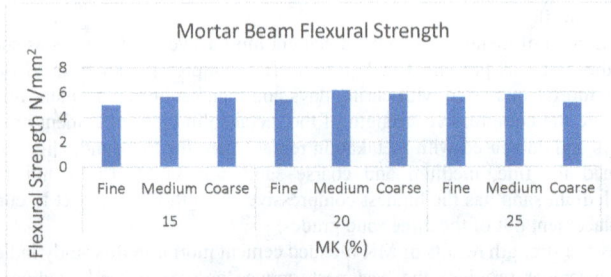

Figure 5: Flexural Strength of MK Blended Cement Mortar with varying Sand Grading and MK

The flexural strength results of MK blended cement mortar showed that 20% MK replacement provides the highest bending strength across all sand grading. At 28 days, medium sand reaches a peak flexural strength of 6.181 N/mm², while fine sand exhibits the lowest, with 5.419 N/mm². Coarse sand shows intermediate performance, achieving 5.854 N/mm². These results are consistent with the ASTM C 618 standards and align with previous studies, such as [20], which demonstrated that 10-20% metakaolin replacement maximizes flexural strength.

The observed differences in flexural strength can be explained by the packing density theory. Medium sand, with its optimal particle size distribution, allows better compaction, reducing voids and enhancing load distribution under bending stress. Conversely, fine sand results in higher water demand and increased porosity, which weakens the mortar's resistance to flexural loads. Futhermore, the trend of flexural strength aligns with that of compressive strength, affirming that 20% metakaolin replacement with medium sand offers the best balance for achieving high flexural strength in mortar, while fine sand presents limitations due to higher porosity and lower density.

Conclusions

From the results of the various tests performed, the following conclusions can be drawn:

i. Calcined kaolin obtained from Ibadan, Oyo state Nigeria exhibits pozzolanic property and could be classified as a Class N Pozzolan.

ii. The fineness test showed that the metakaolin used had an average fineness of 75 μm, which is critical for increasing its reactivity.

iii. The sand was found to be well graded, and its size has significant effects on the properties of cement composite. Therefore, the fine, medium and coarse sand are suitable for making good mortar.

iv. Medium sand has balanced workability, compressive strength and flexural strength, suitable for most general-purpose applications (e.g., concrete, mortar) and moderate shrinkage potential.

v. Blended cement produced with 20% replacement of cement with MK together with medium sand provides mortar with the best balance of workability, water absorption, and strength.

Acknowledgements

The authors acknowledge the following:

• The Tertiary Education Trust Fund (TETFund) for the grant award through the National Research Fund (NRF) 2021 for this research, with research grant award reference number: **TETF/DR&D/CE/NRF2021/SETI/III/00030/VOL.1** without which this research would not have been possible of which part of the research works is hereby published.

• Also worthy of acknowledgement is the leader of LAUTECH Cement and Concrete Research (CEMCON) Group in person of Prof. Akeem Ayinde Raheem for his contribution to ensure

Advances in Cement and Concrete
Materials Research Proceedings 51 (2025) 58-67

Materials Research Forum LLC
https://doi.org/10.21741/9781644903537-7

the successful completion of the research and through whom the research team was encouraged to develop the concept note which eventually earned us the award of the research grant by the NRF.

- The technical support rendered by Engr. S. O. Onadiran, Mr. O. O. Ibikunle and Mr. L. O. Abogunde, all of Civil Engineering Department, LAUTECH Ogbomoso.

References

[1] ACI Committee 211, Standard Practice for Selecting Proportions for Normal, Heavyweight, and Mass Concrete (ACI 211.1-91). American Concrete Institute. (2014)

[2] A. M. Neville, Properties of Concrete. Pearson Education. (2011).

[3] P. K. Metha and P. J. Monteiro, Concrete: Microstructure, Properties, and Materials. McGraw-Hill Education. (2014).

[4] R. Kumar and B. Bhattacharjee, Porosity, Pore Size Distribution and in-Situ Strength of Concrete. Cement and Concrete Research, 33 (2003) 155-164. https://doi.org/10.1016/S0008-8846(02)00942-0. https://doi.org/10.1016/S0008-8846(02)00942-0

[5] L. Bertolini, B. Elsener, P. Pedeferri, and R. B. Polder, Corrosion of Steel in Concrete: Prevention, Diagnosis, Repair. Wiley-VCH. (2004). https://doi.org/10.1002/3527603379. https://doi.org/10.1002/3527603379

[6] G. M. Idorn, Durability of Concrete Structures. CRC Press. (1997).

[7] J. Ambroise and J. Péra, Effects of Metakaolin on the Properties of Concrete. Cement and Concrete Research, 34 (2004) 1521-1528. https://doi.org/10.1016/j.cemconres.2004.04.035

[8] J. M. Justice and T. Ueda, Metakaolin: A Review of Its Use in Concrete. Cement and Concrete Composites, 33 (2011) 881-891. https://doi.org/10.1016/j.cemconcomp.2011.02.011

[9] E. Badogiannis and S. Tsivilis, Exploitation of Poor Kaolin Deposits by Calcination. Cement and Concrete Research, 39 (2009) 1019-1026.

[10] J. M. Shilstone, Concrete Mixture Optimization. Concrete International, 12 (1990) 33-40.

[11] ASTM C618, Standard Specification for Coal Fly Ash and Raw or Calcined Natural Pozzolan for Use in Concrete. (2008).

[12] BS 3148, British Standard Methods of Test for Water for Making Concrete (including Notes on the Suitability of the Water). (1970).

[13] BS EN 196-1, Method of Testing Cement. (1995).

[14] BS1881-116, Method for determination of compressive strength of concrete. (1983).

[15] BS 1881- 118, Method for determination of flexural strength of concrete. (1983).

[16] BS 1881-102, Testing Concrete - Part 102: Method for Determination of Slump. British Standards Institution, London, UK. (1983).

[17] BS 1881-103, Testing Concrete - Part 103: Method for Determination of Compacting Factor. British Standards Institution, London, UK. (1983).

[18] ASTM C618-19, Standard Specification for Coal Fly Ash and Raw or Calcined Natural Pozzolan for Use in Concrete. ASTM International, West Conshohocken, PA. (2019).

[20] S. Wild, J. M. Khatib and A. Jones, Relative strength, pozzolanic activity and cement hydration in superplasticised metakaolin concrete. Cement and Concrete Research, 26 (1996) 1537-1544. https://doi.org/10.1016/0008-8846(96)00148-2

Advances in Cement and Concrete
Materials Research Proceedings 51 (2025) 68-77

Materials Research Forum LLC
https://doi.org/10.21741/9781644903537-8

The use of pozzolanic materials for road stabilization in Africa: A review

Ayodeji Theophilus AKINBULUMA[1,a*], Damilola Ayodele OGUNDARE[2,b] and Ubagaram Johnson ALENGARAM[3,c]

[1]Olusegun Agagu University of Science and Technology, Okitipupa-Igbokoda Road, P.M.B. 345, Okitipupa, Ondo State, 350105 Nigeria

[2]Federal Polytechnic, Ede, Osun State, 232104, Nigeria

[3]Centre for Innovative Construction Technology (CICT) University of Malaya 50603, Kuala Lumpur, Malaysia

[a]at.akinbuluma@oaustech.edu.ng,[b]engrbamidele@gmail.com, [c]johnson@um.edu.my

Keywords: Cementous, Durability, Pozzolanic and Stabilization, Pavement Structure, African Roads

Abstract. The use of pozzolanic materials for road stabilization in Africa has gained significant attention due to the need for sustainable, cost-effective road construction and maintenance solutions. This study reviews the utilization of materials like fly ash and silica fume to enhance the mechanical and durability properties of road pavements in Africa. It highlights challenges such as limited resources, climate variability, and increasing traffic demands while emphasizing the unique properties of pozzolanic materials. These materials react with calcium hydroxide in water to form cementitious compounds, improving pavement strength, reducing permeability, and enhancing resistance to environmental degradation. The study discusses material characterization, mix design optimization, and construction techniques, offering guidance on selecting suitable pozzolanic materials and optimal mix proportions. By leveraging these materials, African nations can enhance road performance, lower maintenance costs, and support sustainable practices. Collaboration among researchers, practitioners, and policymakers is crucial to ensure proper material selection, design, and quality control.

Introduction

Roads provide vital links for trade, travel, and social integration, road infrastructure is crucial for Africa's economic development [1]. However, because traditional construction materials are expensive and the continent faces challenging environmental conditions, including seasonality in climate change and low soil quality in many parts, building and maintaining durable roads is a major challenge. Pozzolanic materials provide solution to these problems when it comes to stabilizing roads.

A cementitious compound that can bind other materials is created when pozzolanic materials which can be either siliceous or siliceous and aluminous combine with lime and water (Fig. 1). These materials include calcined clays, fly ash, and slag from industry, as well as natural pozzolans like volcanic ash amongst others. Pozzolanic materials have been used since antiquity, most notably in Roman concrete, which is still considered the gold standard for durability [2]. Pozzolanic minerals include illite, mica, kaolinite, and montmorillonite. Heat treatment of natural materials containing pozzolanas, such as clays, shales, and some silicious rocks, produces artificial pozzolanas, such as ashes. When plants are burnt, silica from the soil is removed as nutrients and is left behind in the ashes, which adds to the pozzolanic element. Rich in silica, rice husk ash, rice straw, and bagasse combine to create an excellent pozzolana [3].

Advances in Cement and Concrete | Materials Research Forum LLC
Materials Research Proceedings 51 (2025) 68-77 | https://doi.org/10.21741/9781644903537-8

Three primary types of pavements are built in African countries: unpaved (gravel/earth), semi-rigid, and flexible. Over 95% of paved roads are made of flexible pavement, which is applied over a granular, unbound, or treated (chemically stabilized) base course and has either surface treatment or bituminous mix for a surface course. Penetration macadam is a base course for road that can withstand 3 to 10 million ESAL [4]. According to the Pavement Design and Materials Manual, a subgrade with at least 15% 4-day soaked CBR should be placed on top of a 50mm asphalt concrete surface course, 150mm crushed rock granular foundation, and 200mm subbase made of gravel or soil with a minimum of 20% 4-day soaked CBR at 95% Maximum Dry Density (MDD) [5].

The Manual suggests using clean, unweather boulders with a minimum diameter of 0.3 meters and no soil fines, or sturdy crushed rock with a minimum Ten Percent Fines Value (TFV) soaked = 75% TFVdry and a Ten Percent Fines Value of 110KN for the granular base. In the coastal regions of some African countries like Nigeria, Mali, Burkina Faso, Ghana, Niger amongst others, with abundance of limestone aggregates, it can be challenging to meet the requirements for granular base course material. These regions have significant building costs because quality materials for roads must be transported across great distances. Research into the potential use of naturally occurring pozzolans for road construction in numerous African nations, including Tanzania and Nigeria, among others, was motivated by the use of granulated blast furnace slag on the Dar-Mlandizi Road project [7]. This paper therefore examines the effective use of pozzolanic materials for road stabilization in Africa.

Figure 1: Pozzolanic Materials [6]

Challenges in the Construction and Maintenance of Roads in Africa

The road networks in Africa, largely inherited from colonial administrations, were constructed to fulfill limited purposes, such as facilitating resource extraction and connecting administrative hubs. Consequently, these networks have proven inadequate in meeting the transportation needs of independent African states. Post-independence maintenance and expansion efforts have faced significant challenges, leading to low road density and deteriorating infrastructure. Rural access remains particularly poor, with approximately two-thirds of the rural population unable to access all-weather roads. Urban areas face their own challenges as rapid population growth outpaces infrastructure development. Furthermore, climate change exacerbates existing vulnerabilities through extreme weather events and land degradation, while chronic underfunding undermines the maintenance and improvement of existing networks. This section analyzes strategies implemented across various African regions to address these challenges and examines the limitations of these approaches [8][9].

North African Countries: Egypt Strategy

Egypt has implemented the National Roads Project to enhance its transportation infrastructure, allocating 0.7% of GDP to the sector. However, a disproportionately low share—approximately

Advances in Cement and Concrete
Materials Research Proceedings 51 (2025) 68-77

Materials Research Forum LLC
https://doi.org/10.21741/9781644903537-8

0.15%—is directed toward road maintenance. This inadequate allocation has contributed to widespread infrastructure deterioration, leading to frequent traffic accidents, which collectively impose a financial burden equivalent to 1.5% of GDP [10]. Maintenance of the national road network is managed by the General Authority of Roads, Bridges, and Land Transport (GARBLT). Despite government efforts to modernize and maintain the road system, significant challenges persist due to limited technical capacity and insufficient funding mechanisms. Current data indicate that 75% of roads, along with numerous bridges, are in a state of disrepair, adversely affecting both domestic mobility and international transportation connectivity [11]. Overall, there is still a problem with the way Egypt's road network is maintained (Table 1).

Central African Nations: Angola Strategy
Angola's road infrastructure, comprising 72,323 km of roads, is in dire need of rehabilitation following years of neglect due to protracted civil conflict. The government allocates approximately 14% of its GDP annually to infrastructure development, with a significant portion directed toward road rehabilitation and construction. The National Roads Institute of Angola (INEA) oversees maintenance activities; however, the sustainability of these efforts is undermined by reliance on state budgets and the underutilization of the proposed Road Fund (RF). To address funding gaps, INEA has recommended reforms such as increased user charges and adjustments to tax policies. While substantial progress has been achieved in reconstructing the road network, Angola must implement long-term strategies to ensure the viability and sustainability of its infrastructure development initiatives [12].

Southern African Countries: South Africa Strategy
South Africa boasts the most extensive road network on the continent, spanning 746,978 km, with 153,719 km paved. The South African National Roads Agency Limited (SANRAL) is tasked with maintaining the national road network and employs innovative strategies such as performance-based contracting and partnerships with private sector entities. Despite these efforts, approximately 38% of roads are classified as being in fair to poor condition due to inadequate maintenance at the municipal level and a substantial financial backlog estimated at 80–149 billion Rand. South Africa's toll-road system and concession contracts are considered effective models for infrastructure financing and management. However, persistent funding shortages and data limitations within municipalities present ongoing challenges to achieving sustainable road maintenance.

East African Countries: Ethiopia Strategy
Ethiopia has made significant strides in expanding its road infrastructure, increasing the total network length from 6,400 km in 1951 to 85,966 km by 2015. This progress is attributed to reforms in road management and financing, spearheaded by the Ethiopian Roads Authority (ERA) and the establishment of the Road Fund in 1997. The Road Fund operates on a fee-for-service model, which has proven effective in ensuring sustainable financing for road maintenance. These institutional and financial reforms have fostered a culture of efficient management and timely maintenance, positioning Ethiopia as a regional leader in road infrastructure development. However, as the network continues to expand, challenges related to capacity building, technological adoption, and climate resilience must be addressed to sustain this progress [13][14][15].

West African Countries: Nigeria Strategy
Nigeria possesses the largest road network in West Africa, yet it is plagued by chronic maintenance challenges. Corruption, insufficient funding, and reliance on outdated construction practices have contributed to the poor state of the country's roads. The Federal Roads Maintenance Agency (FERMA) and state ministries of works are responsible for road maintenance, but their efforts are

Advances in Cement and Concrete
Materials Research Proceedings 51 (2025) 68-77

Materials Research Forum LLC
https://doi.org/10.21741/9781644903537-8

often undermined by inefficiency and resource constraints. Unlike Ethiopia, which has implemented sustainable financing mechanisms, Nigeria has struggled to establish effective frameworks for funding and managing its road infrastructure. The resulting delays in maintenance and construction have hindered economic growth and development, highlighting the urgent need for institutional reforms and innovative financing solutions [16][17][18].

Table 1: Comparison Between Road Maintenance Perspectives of the Studied African Countries

Country	Road Fund	Maintenance Culture	Main Challenges
Egypt	▪ Annual fund from the government	Very poor	▪ Inadequate funding ▪ Lack of information
Angola	▪ Annual fund from the government ▪ Road fund model (partial)	Poor	▪ Lack of Information ▪ Inadequate fund needed to develop network
South Africa	▪ Annual fund from the government ▪ Road user charge model	Fair	▪ Poor data at municipalities
Ethiopia	▪ Road fund model	Good	▪ Insufficient funding
Nigeria	▪ Annual Fund from the government	Very Poor	▪ Insufficient funding ▪ Lack of proper administration ▪ Corruption

Unique Properties of Pozzolanic Materials that make them Ideal for Road Stabilization

Pozzolanic materials have gained increased attention due to the civil engineering community's demand for sustainable and cost-effective solutions, particularly for road stabilization. These materials, such as fly ash, volcanic ash, and silica fume, offer unique chemical and physical properties that make them effective alternatives to conventional stabilizers like lime and cement [19]. Construction engineers have long used these materials for various applications, including road stabilization, with their usage becoming more prominent as the need for sustainable infrastructure intensifies. For example, the ability of pozzolanic materials to react with calcium hydroxide and form calcium silicate hydrate (C-S-H) improves soil binding and reduces porosity, enhancing the durability of stabilized layers and ensuring longer-lasting road pavements [20]. Their utilization aligns with global trends in sustainable infrastructure development, as illustrated in Figure 2, which compares the properties of different stabilizing materials, such as volcanic ash processed under various conditions.

Incorporating pozzolanic materials into soil stabilization significantly improves the mechanical properties of roads. Enhanced unconfined compressive strength (UCS) ensures better load-bearing capacity, making roads more resistant to deformation caused by traffic loads [21]. The pozzolanic reaction also reduces soil permeability, mitigating the effects of moisture infiltration that weaken conventional road bases over time. These benefits lead to improved durability and performance of road infrastructure, particularly in regions prone to adverse environmental conditions. Moreover, pozzolanic materials contribute to cost-effectiveness, as they are often more affordable than traditional stabilizers. The abundance of natural pozzolans, such as volcanic ash in regions across Africa, combined with lower transportation and processing costs, makes them an economically viable option [22]. Industrial byproducts like rice husk ash and fly ash further reduce costs while simultaneously addressing waste management issues, highlighting their dual role in economic efficiency and environmental sustainability [23].

Beyond cost and mechanical improvements, pozzolanic materials enhance chemical durability, offering resistance to sulfate attack, a common challenge in road stabilization. Sulfate-rich soils often degrade lime- or cement-stabilized layers, whereas pozzolanic materials, due to their lower calcium content and the formation of C-S-H, are less susceptible to such degradation. This makes them particularly suitable for arid and semi-arid regions, where sulfate concentrations in the soil are higher [24]. Additionally, the fine particle size of pozzolanic materials improves soil workability during stabilization, enabling uniform distribution and cohesion within the soil matrix [25]. Engineers can tailor the stabilization process to meet specific project needs, ensuring optimal load-bearing capacity, reduced shrinkage, or environmental adaptability. These properties enhance construction flexibility, ensuring that pozzolanic-stabilized roads meet both structural and environmental demands [26].

The environmental benefits of using pozzolanic materials in road construction are substantial. Many pozzolanic materials, including fly ash and silica fume, are industrial byproducts that reduce waste by repurposing materials otherwise destined for disposal. This approach lowers the environmental footprint of road construction while also decreasing the reliance on energy-intensive cement production, a significant source of carbon dioxide emissions [27]. Figure 3 illustrates the pozzolanic reaction, which complements cement hydration to improve soil stabilization. Furthermore, material characterization, mix design optimization, and proper construction techniques are critical to the successful implementation of pozzolanic materials in road stabilization. These steps ensure that materials meet necessary performance standards, with iterative testing and quality control ensuring uniformity and long-term reliability of stabilized pavements [28-29].

*Figure 2: Stabilizing Materials for Road Construction (Cement (C); fine volcanic ash (VA); ultra-fine volcanic ash (VAF); and VA at 550 °C (VA550), VA at 650 °C (VA650), and VA at 750 °C (VA750)), **Source:** Khan et al, (2022)*

Advances in Cement and Concrete
Materials Research Proceedings 51 (2025) 68-77

Materials Research Forum LLC
https://doi.org/10.21741/9781644903537-8

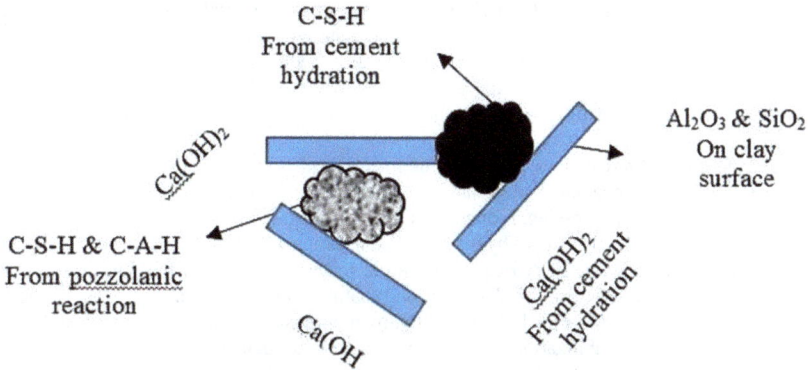

Figure 3: *Pozzolanic Reaction and Cement Hydration [36]*

Economic benefits of using locally available pozzolanic materials for road stabilization in Africa

Africa faces the challenge of expanding its road infrastructure to support economic growth while ensuring cost-effectiveness and sustainability. The use of locally available pozzolanic materials, such as volcanic ash, fly ash, and rice husk ash, offers substantial economic benefits for road stabilization. One key advantage is the potential to reduce construction costs. Traditional stabilizing agents like cement and lime are often expensive to import, especially in remote areas with limited transportation infrastructure. By sourcing pozzolanic materials locally, transportation costs are minimized, and the reliance on imported materials is reduced. This approach is especially beneficial in rural and underdeveloped regions, where budgets for road construction are constrained, as noted by [30].

Africa's natural abundance of pozzolanic resources, including volcanic ash found in Kenya, Tanzania, and Ethiopia, provides opportunities for resource utilization and local economic development. By leveraging these materials, countries can decrease their dependence on imported products and stimulate local industries associated with extraction and processing. This not only supports job creation but also aligns with broader regional development goals by promoting sustainable practices [31]. Additionally, the use of locally sourced materials contributes to long-term economic growth by fostering industries and enhancing local expertise in material processing and application.

The durability of pozzolanic-stabilized roads contributes to long-term cost savings by reducing the frequency and costs of repairs. These materials enhance the resilience of roads to environmental factors such as moisture, temperature fluctuations, and chemical exposure, ensuring a longer lifespan and minimizing lifecycle costs. For governments and communities with limited resources, these savings are critical for sustaining road networks. Furthermore, the adoption of pozzolanic materials supports sustainable construction practices by repurposing industrial byproducts and reducing the environmental impact of material extraction. This alignment with global sustainability goals enhances the reputation of local construction industries and can attract international funding and investments for infrastructure development [32, 33].

The Environmental Sustainability Impact of Using Pozzolanic Materials for Road Stabilization in Africa

The need for environmentally sustainable road construction in Africa is increasingly critical due to rapid urbanization and economic growth. Traditional road stabilization methods, heavily reliant on cement and lime, contribute to environmental degradation through resource extraction and significant carbon emissions. In contrast, pozzolanic materials, such as fly ash, volcanic ash, and rice husk ash, present an alternative that significantly reduces the carbon footprint of road construction. Cement production, a major source of global carbon dioxide emissions, can be partially or fully replaced by pozzolanic materials, contributing to climate change mitigation efforts [4].

The environmental advantages of pozzolanic materials also include minimizing the extraction of natural resources. Pozzolanic materials, often industrial byproducts or naturally occurring substances, require less processing, reducing the environmental impact associated with quarrying and mining. For example, fly ash is a byproduct of coal combustion, while volcanic ash is naturally abundant in specific regions of Africa. Using these materials decreases the need for new raw material extraction, preserving biodiversity and natural landscapes, as well as supporting sustainable development goals [34]. Additionally, by repurposing industrial byproducts like silica fume and fly ash, road stabilization practices promote waste reduction and resource efficiency. This approach diverts waste from landfills, transforming it into a valuable resource for infrastructure development [35].

Improved soil and water quality is another environmental benefit of using pozzolanic materials for road stabilization. Unlike cement and lime, which can lead to soil alkalinization and water runoff contamination, pozzolanic materials tend to form stable compounds with minimal environmental impact. Their application reduces leaching risks and safeguards surrounding ecosystems, particularly in regions where agriculture or natural habitats coexist with infrastructure development [36]. Furthermore, pozzolanic materials contribute to the durability of roads, reducing the frequency of reconstruction and minimizing resource consumption over time. The improved resilience of pozzolanic-stabilized roads to extreme weather events, such as flooding, enhances climate adaptation and reduces the environmental disruptions associated with repair activities.

Conclusion

The study has comprehensively explored the potential of pozzolanic materials for enhancing road stability in Africa, assessing their suitability, environmental benefits, and effectiveness in different soil types; as well as seeks to identify best practices, address challenges, and provide recommendations for sustainable, cost-effective road construction across the African continent. This was carried out qualitatively. Therefore, based on the reviewed literature, conclusions were made as follows:

- Pozzolanic materials offer a viable solution to the challenges of road durability, cost-efficiency, and environmental sustainability in Africa.
- When combined with traditional binders like lime or cement, materials such as volcanic ash, calcined clays, and industrial by-products like fly ash, it provides increased strength and longevity to road pavements.
- These materials are particularly effective in regions with expansive soils or areas prone to erosion, such as Nigeria, South Africa, and Angola, where conventional methods may fail.
- The adoption of pozzolanic materials can significantly reduce the carbon footprint associated with road construction, supporting global sustainability goals.
- For widespread application to succeed, challenges related to the availability, quality control, and technical expertise of using pozzolanic materials must be addressed.

Advances in Cement and Concrete   Materials Research Forum LLC
Materials Research Proceedings 51 (2025) 68-77   https://doi.org/10.21741/9781644903537-8

References

[1] D. G. Owusu-Manu, A. B. Jehuri, D. J. Edwards, F. Boateng, and G. Asumadu, The impact of infrastructure development on economic growth in sub-Saharan Africa with special focus on Ghana. Journal of Financial Management of Property and Construction. 3 (2019) 253-273. https://doi.org/10.1108/JFMPC-09-2018-0050

[2] K. Shen, X. Qian, C. Hu, and F. Wang, Revisiting Ancient Roman Cement: The Environmental-Friendly Cementitious Material Using Calcium Hydroxide-Sodium Sulfate-Calcined Clay. ACS Sustainable Chemistry & Engineering. 13 (2023) 5164-5174 https://doi.org/10.1021/acssuschemeng.2c07495

[3] D. Nawir, M. D Bakri, and I. A. Syarif, Central government role in road infrastructure development and economic growth in the form of future study: the case of Indonesia. City, Territory and Architecture. 1 (2023) 12 - 22. https://doi.org/10.1186/s40410-022-00188-9

[4] S. Kumar, Design and Construction of Bituminous Pavements. Journal of Mechanical and Construction Engineering (JMCE). 2 (2 022) 1-16. https://doi.org/10.54060/jmce.v2i2.19

[5] C. Wang, M. K. Lim, X. Zhang, L. Zhao, and P. T. W. Lee, Railway and road infrastructure in the Belt and Road Initiative countries: Estimating the impact of transport infrastructure on economic growth. Transportation Research Part A: Policy and Practice. 134 (2020) 288-307. https://doi.org/10.1016/j.tra.2020.02.009

[6] L. H. Ali, and Y. K. Atemimi, Effective Use of Pozzolanic Materials for Stabilizing Expansive Soils: A Review. In IOP Conference Series: Earth and Environmental Science. 1 (2024) 1374. https://doi.org/10.1088/1755-1315/1374/1/012014

[7] N. T. Sithole, and T. Mashifana, Geosynthesis of building and construction materials through alkaline activation of granulated blast furnace slag. Construction and Building Materials, 13 (2020) 264 -273 https://doi.org/10.1016/j.conbuildmat.2020.120712

[8] M. F. Murove, and M. F. Murove, Regional Integration and the Ethic of Co-operation in Post-colonial Africa. African Politics and Ethics: Exploring New Dimensions. 16 (2020) 115-138. https://doi.org/10.1007/978-3-030-54185-9_7

[9] M. Bi, and Z. Zhang, Exploring the path of autonomous development: The development dilemma and coping strategies of Sub-Saharan Africa in the post-epidemic era. Journal of the Knowledge Economy. 15 (2024) 5043-5071. https://doi.org/10.1007/s13132-023-01145-8

[10] M. B. Bouraima, Y. Qiu, B. Yusupov, and C. M. Ndjegwes, A study on the development strategy of the railway transportation system in the West African Economic and Monetary Union (WAEMU) based on the SWOT/AHP technique. Scientific African. 8 (2020) 388. https://doi.org/10.1016/j.sciaf.2020.e00388

[11] R. Khallaf, J. Guevara, P. Mendez-Gonzalez, and G. Castelblanco, A system dynamics model for a national PPP program: The Egyptian project portfolio. In Construction Research Congress 11 (2024) 507-516 https://doi.org/10.1061/9780784485286.051

[12] H. M. Mostafa, Road maintenance in Africa: approaches and perspectives. In E3S Web of Conferences EDP Sciences. 38 (2018) 73 - 81 https://doi.org/10.1051/e3sconf/20183801005

[13] A. Ewnetu, The Effect of Rural Road Access on Rural Households Livelihood Improvement: Evidence from Selected Weredas in Amhara Regional State, Ethiopia. Journal of African Development Studies. 1 (2023) 47-63. https://doi.org/10.56302/jads.v10i1.8780

[14] M. G. Sisay, The pitfalls of Ethiopian road developments: socio-economic impacts. Cogent Social Sciences. 10 (2024) 2319220. https://doi.org/10.1080/23311886.2024.2319220

[15] L. M. Hailemariam, and D. A. Nuramo, Factor analysis of key parameters for effective design delivery of urban transport infrastructure in Ethiopia. Heliyon. 17 (2024) 89-102 https://doi.org/10.1016/j.heliyon.2024.e34681

[16] S. I. Oni, and K. Ojekunle, Transportation and agro-food distribution in sub-saharan africa. Lagos Journal of Geographical Issues. 3(2023) 75-89.

[17] O. Anjorin, Spatial analysis of transportation infrastructure distribution in Adamawa State, Nigeria: A location quotient perspective. Journal of the Bulgarian Geographical Society. 50 (2024) 113-128. https://doi.org/10.3897/jbgs.e115392

[18] O. O. Adepoju, Analysis of road transportation infrastructure construction and maintenance for sustainable development in South-Western Nigeria. Journal of Sustainable Development of Transport and Logistics. 1 (2021) 49-58. https://doi.org/10.14254/jsdtl.2021.6-1.4

[19] O. A. Mohamed, A. A. Farghali, A. K. Eessaa, and A. M. El-Shamy, Cost-effective and green additives of pozzolanic material derived from the waste of alum sludge for successful replacement of portland cement. Scientific Reports. 12 (2022) 20974. https://doi.org/10.1038/s41598-022-25246-7

[20] K. Weise, N. Ukrainczyk, and E. Koenders, Pozzolanic reactions of metakaolin with calcium hydroxide: review on hydrate phase formations and effect of alkali hydroxides, carbonates and sulfates. Materials & Design. 231 (2023) 112062. https://doi.org/10.1016/j.matdes.2023.112062

[21] D. T. Nguyen, and V. T. A. Phan, Engineering properties of soil stabilized with cement and fly ash for sustainable road construction. International Journal of Engineering, Transactions C: Aspects. 12 (2021) 2665-2671.

[22] H. Abdurrahman, N. Rizaldi, M. F. Wijaya, M. Olivia, and G. Wibisono, Utilization of Pozzolanic Material to Improve the Mechanical Properties of Crumb Rubber Concrete as Rigid Pavement-A Review. In Journal of Physics: Conference Series. 2049 (2021) 012087 https://doi.org/10.1088/1742-6596/2049/1/012087

[23] J. Lu, S. Y. Jiang, J. Chen, C. H. Lee, Z. Cai, and H. D. Ruan, Fabrication of superhydrophobic soil stabilizers derived from solid wastes applied for road construction: A review. Transportation Geotechnics. 40 (2023) 100974. https://doi.org/10.1016/j.trgeo.2023.100974

[24] N. S. Jama, and K. A. Saeed, Pozzolanic materials for stabilization/solidification of soil contaminated by heavy metals-a review. Journal of Engineering and Sustainable Development. 4 (2023) 487-498. https://doi.org/10.31272/jeasd.27.4.6

[25] T. Sengul, N. Akray, and Y. Vitosoglu, Investigating the effects of stabilization carried out using fly ash and polypropylene fiber on the properties of highway clay soils. Construction and Building Materials. 400 (2023) 132590. https://doi.org/10.1016/j.conbuildmat.2023.132590

[26] L. S. Wong, R. Hashim, and F. Ali, Improved strength and reduced permeability of stabilized peat: Focus on application of kaolin as a pozzolanic additive. Construction and Building Materials. (2013). 40, 783-792. https://doi.org/10.1016/j.conbuildmat.2012.11.065

[27] U. Zada, A. Jamal, M. Iqbal, S. M. Eldin, M. Almoshaogeh, S. R. Bekkouche, and S. Almuaythir, Recent advances in expansive soil stabilization using admixtures: current challenges

Advances in Cement and Concrete
Materials Research Proceedings 51 (2025) 68-77

Materials Research Forum LLC
https://doi.org/10.21741/9781644903537-8

and opportunities. Case Studies in Construction Materials. 18 (2023) 01985. https://doi.org/10.1016/j.cscm.2023.e01985

[28] N. S. Jama, and K. A. Saeed, Pozzolanic materials for stabilization/solidification of soil contaminated by heavy metals-a review. Journal of Engineering and Sustainable Development. 4 (2023) 487-498. https://doi.org/10.31272/jeasd.27.4.6

[29] P. P. Kulkarni, and J. N. Mandal, Strength evaluation of soil stabilized with nano silica-cement mixes as road construction material. Construction and Building Materials. 314 (2022) 125363. https://doi.org/10.1016/j.conbuildmat.2021.125363

[30] M. Amran, A. M. Onaizi, N. Makul, H. S. Abdelgader, W. C. Tang, B. T. Alsulami, and Y. Gamil, Shrinkage mitigation in alkali-activated composites: A comprehensive insight into the potential applications for sustainable construction. Results in Engineering. 18 (2023) 101452. https://doi.org/10.1016/j.rineng.2023.101452

[31] K. Y. W. N'Simba, J. W. Kaluli, B. M. Mwangi, and H. M. Mwangi, Effect of thermal treatment on the pozzolanic activity of natural clay from selected Kenyan wetlands. Journal of Sustainable Research in Engineering. 4 (2023) 137-154.

[32] M. Cavalieri, P. L. Ferrara, C. Finocchiaro, and M. F. Martorana, An Economic Analysis of the Use of Local Natural Waste: Volcanic Ash of Mt. Etna Volcano (Italy) for Geopolymer Production. Sustainability. 2 (2024) 740-756 https://doi.org/10.3390/su16020740

[33] Z. Vincevica-Gaile, T. Teppand, M. Kriipsalu, M. Krievans, Y. Jani, M. Klavins, and J. Burlakovs, Towards sustainable soil stabilization in peatlands: Secondary raw materials as an alternative. Sustainability. 12 (2021) 6726. https://doi.org/10.3390/su13126726

[34] A. A. Mohammed, H. Nahazanan, N. A. M. Nasir, G. F. Huseien, and A. H. Saad, Calcium-based binders in concrete or soil stabilization: challenges, problems, and calcined clay as partial replacement to produce low-carbon cement. Materials. 5 (2023) 486-503 https://doi.org/10.3390/ma16052020

Advances in Cement and Concrete
Materials Research Proceedings 51 (2025) 78-84

Materials Research Forum LLC
https://doi.org/10.21741/9781644903537-9

Waste paper sludge ash used as replacement for cement in concrete production

S.O. NWAUBANI[1,a*], I.I. OBIANYO[2b] and P.A. ONWUALU[3,c]

[1]African University of Science and Technology, Abuja, Nigeria

[1]School of Civil and Environmental Engineering, University of the Witwatersrand, South Africa

[2]Nile University of Nigeria, Abuja, Nigeria

[3]African University of Science and Technology, Abuja, Nigeria

[a]snwaubani@aust.edu.ng, [b]ifeyinwa.obianyo@nileuniversity.edu.ng, [c]aonwualu@aust.edu.ng

Keywords: Waste Paper Sludge, Cement Replacement, Eco-Friendly Concrete, Pozzolanic Material, WPSA

Abstract. The use of waste is rapidly becoming a newly emerging supra-disciplinary field in most parts of the world where the use of industrial wastes like fly ash, granulated steel slag, silica fume, waste fibers etc., in construction has become very popular since the last half of the 20th century. However, lack of standards and the existing concept of trying to match or possibly exceed the strength of a control mix made with 100% Portland cement is hampering the increased use of waste materials in construction. Waste paper mill sludge is a major economic and environmental problem for the paper and board industry. Enormous quantities of waste paper sludge are generated all around the world. The material is a by-product of the de-inking and re-pulping of paper during the production of new papers, boards and recycled papers. The main recycling and disposal processes for the waste paper sludge are either as agricultural fertilizer where the sludge is dried and spread directly over agricultural land for its lime content, incineration at the paper mills as a source of heat for the mill's boilers, or simply transported and dumped onto public or private landfills. This study looked at the chemical, physical, and mineralogical characterization of the waste paper sludge ash (WPSA), from a South African source. The influence of the waste paper sludge ash on concrete incorporating 5%, 10%, 20%, 30% and 50% of the waste sludge were investigated with a constant water to binder ratio of 0.5. The results of the investigation conclude that WPSA is a viable cement extender and has potential for use in the construction industry.

Introduction

Waste paper sludge ash (WPSA) contains reactive silica and alumina, as well as lime (CaO) that are similar to the mineral compounds found in Portland cement. It could therefore be a suitable cementitious material since it provides a source of additional silica and alumina. Consequently, new emerging technologies and research worldwide are now focussed on using waste paper sludge ash in concrete construction to replace part of the Portland cement.

Although very little work has been done in Africa on the research of paper pulp ash as a viable cement replacement material, researchers in India and some countries within the European Union have embarked on this study extensively [1, 2, 3]. Several experimental studies have confirmed that WPSA is a substitute cementitious material mainly due to the presence of CaO, SiO_2, Al_2O_3 and Fe_2O_3. These chemical characteristics are similar to those in OPC and ensure the formation of extra hydration products which densifies the microstructure and improves durability performance.

Advances in Cement and Concrete
Materials Research Proceedings 51 (2025) 78-84

Materials Research Forum LLC
https://doi.org/10.21741/9781644903537-9

Effect on Fresh Concrete

Many of the studies reported involve the replacement of up to 30% of cement with waste paper pulp ash. Due to the high surface area and porosity, mixtures incorporating WPSA have a high-water requirement. Consequently, when a higher amount of paper pulp was included in the mixture, it requires more water to achieve a given slump. Studies reported by Nwaubani [4] and by Kakora and Nwaubani [5], established that the consistency of concrete containing WPSA can only be improved by the use of super-plasticizing admixtures or by the addition of excessive water. Their work clearly showed that the former is the only rational way of achieving adequate workability and higher substitution levels without jeopardizing strength development [5]. Balwaik and his colleagues [6], replaced 5% to 20% by weight of cement mix designed to give 28-day characteristic compressive strengths of 20 MPa and 30MPa respectively. They found that the slump decreased as the paper pulp content in the mixtures increased.

Effect on Hardened Concrete

Baia et al [7], investigated the compressive strength and workability of concrete using WPSA and Ground Granulated Blast Furnace Slag (GGBS). Their paper concluded that it is possible to combine WPSA and GGBS (both industrial by-products), to produce a binder without adding Portland cement. The study and other studies by Sathish et al [8], suggest that resultant concrete was adequate for low strength uses and the compressive strength did not exceed 15 MPa. Ayatollah et al [9], researched the properties of WPSA concrete and their results showed that, 5% replacement levels of OPC with WPSA are optimal for the best strength gain over 28 days. Rashad [10], concluded from his research with WPSA that regardless of the water/binder ratio, the optimal cement replacement is 20% with the addition of heat reducers and superplasticizers. He believed that the effective thermal treatment of the WPSA would give the optimum conditions, subject to factors such as calcination temperature and heating period [10]. In London, Mavroughlidou, Bouzouki [11,12], concluded that when used in modest amounts, WPSA can maintain or improve the concrete strength and water absorption compared to regular CEM-I concrete. However, they did not specify the percentage replacement at the conclusion of their study. Balwaik and his colleagues [6], found that the splitting tensile, compressive and flexural strengths of concrete increased up to 10% on the addition of WPSA but that further increase of WPSA reduces the strengths progressively. It was concluded that 5 to 10 % replacement of waste paper pulp for cement is the most suitable mix proportion [8]. Corinaldesi et al. [13], found that with 5% of WPSA substitution, mortars exhibited higher compressive strength at 28 days than mortars made with OPC only. They also concluded that the mechanical properties of concrete show a positive effect if cement is replaced by less than 10% WPSA. They further noted that if the concrete is to be made by a partial replacement higher than 25%, the concrete should only be used for non-structural elements like pedestrian paving blocks [11], Frias [14], researched the manufacture of binary and ternary blends incorporating WPSA and recommended that the substitution level be limited to 10% replacement for binary blends. He also reported a reduction in the setting times, loss of workability and excessive drying shrinkage during his research. He concluded that the percentage of clinker replaced by the addition of these minerals should not exceed 21% so as to ensure that the workability of the mixture is not adversely affected.

Characterization of Materials Used

Physical Characteristics: Paper sludge described above consists of mineral fillers, small cellulose fibres, water, inorganic salts and organic compounds. The moisture content varies but can be as high as 40%. The material is viscous, sticky and hard to dry and can vary in viscosity and lumpiness. Fig1a and 1b show the physical characteristics of virgin waste paper sludge (WPS) after drying. Fig.1c shows the Scanning Electron Micrograph (SEM), of the material which indicate the presence of irregular pores and fibrous characteristics of the material.

Advances in Cement and Concrete
Materials Research Proceedings 51 (2025) 78-84

Materials Research Forum LLC
https://doi.org/10.21741/9781644903537-9

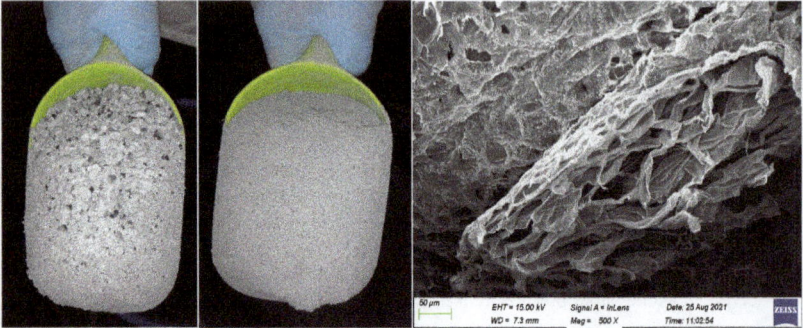

Fig1a: Uncrushed WPA   Fig1b: Crushed WPA   Fig 1c: SEM of Virgin Paper Sludge

The material passing through the 75µm sieve was used in the concrete mixes for this investigation. The particle size distribution (PSD) showed 10% of the sample passing 0.874 μm, 50% passing 2.217 μm, 90% passing of 9.922 μm and 100% passing 10.014 µm.

Thermogravimetric Analysis (TGA): This analysis was carried out using the dried powder material, at a rate of 5°C per minute from a temperature of 100°C up to 800°C. Fig.2 indicate that thermal degradation occurred in the temperature range between 300°C and 400°C, being characterized by the degradation of cellulose and hemicellulose, which decomposes between 320°C and 380°C [15].

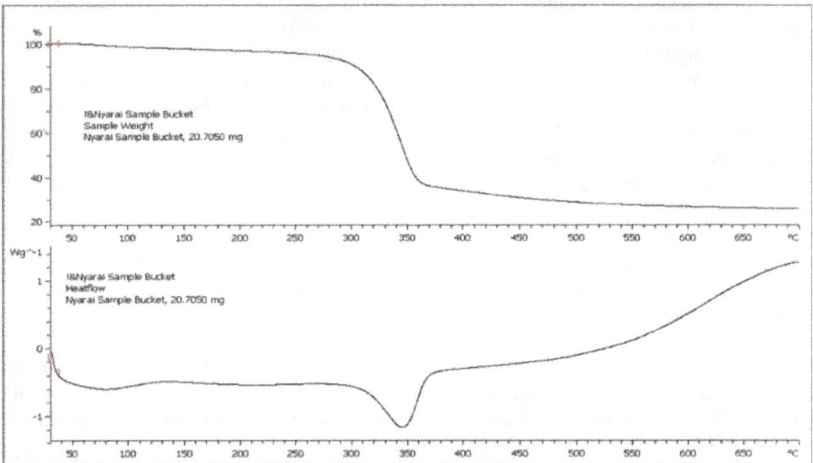

Fig.2: TGA of Paper sludge

Chemical Characteristics of cement and the Paper Sludge Ash

Table 1 below compares the chemical Oxide composition of the Ordinary Portland cement (OPC), used and that of the raw waste paper sludge (WPS) and the calcined material at temperature of 800°C for 2 hours – also referred to here as Waste Paper Sludge Ash (WPSA).

Advances in Cement and Concrete

Materials Research Proceedings 51 (2025) 78-84

Materials Research Forum LLC

https://doi.org/10.21741/9781644903537-9

Table 1: Comparisons of the Physiochemical Properties of Cement, Waste Paper Sludge Ash (DWPS) and WPS.

Description	% by Weight of Cement	% by Weight of WPS	% by Weight of WPSA
SiO_2	20.78	33.54	40.45
Al_2O_3	5.11	5.54	9.20
CaO	60.89	12.82	25.69
MgO	3.00	2.14	2.57
Fe_2O	3.17	2.74	2.69
K_2O	-	0.75	0.98
Na_2O	-	1.40	1.39
SO_3	-	1.47	1.45
LOI	1.71	10.66	0.78
Specific gravity	3.15	2.14	2.54
Moisture content	-	3.46	1.15
Dry density (kg/m^3)	-	690	807

During incineration, organic compounds are burned at between 350 to 800°C, whereas mineral fillers and inorganic salts are transformed into the corresponding oxides at higher temperatures (>800°C). Fig. 2 shows that CaO, Al_2O_3, MgO, and SiO_2 are the most abundant oxides in WPSA. The amounts of the other elements are usually low and depend mostly on the particular mill and the type of paper being produced. Compounds that may be present in small concentrations of less than 2% include: Ferric trioxide (Fe_2O_3), Sulphate (SO_3), Potassium oxide (K_2O), Sodium Oxide (Na_2O), etc. Evidently, these chemical characteristics are similar to those in OPC and ensure the formation of extra hydration products which densifies the microstructure and improves durability performance

It should be noted that depending on the source of the materials, waste paper sludge may be mixed with kaolin clays. Pera and Amrouz [10], studied the production of metakaolin by incinerating paper sludge in the temperature range of 700°C to 800°C for 3 hours and it was shown that kaolinite mineral in the sludge converted to metakaolin. However, no metakaolin was detected in the waste paper sludge when combustion is at temperatures higher than 800°C or for incineration exceeding 5 hours [15].

X-Ray Diffraction (XRD) Analysis

XRD Analysis carried out at Wits University have identified the main compounds in PSA as Gehlenite ($Ca_2Al_2SiO_7$); Free Lime (CaO); Calcite ($CaCO_3$); Quartz (SiO_2), and Merwanite ($Ca_3Mg(SiO_4)_3$).

Advances in Cement and Concrete
Materials Research Proceedings 51 (2025) 78-84

Materials Research Forum LLC
https://doi.org/10.21741/9781644903537-9

Experimental Details

Introduction: The focus of the experimental programme is to determine whether WPSA is a viable OPC replacement. This section presents the chosen mix designs for concrete mixes with OPC replacement levels of 5%, 10%, 20%, 30% and 50% by mass. The influence of WPSA addition on the fresh and hardened properties of the concrete were studied. Slump tests were used to determine the fresh characteristics (workability) of the concrete while compressive strength tests and durability index tests were done to determine the hardened characteristics of the concrete.

Mix Design

The mix designs were done according to the Concrete & Cement Institute (C&CI) design method which is similar to the ACI Standard 221.1-91. For this method, the optimum mix quantities depend on the stone size, the compacted bulk density (CBD) of the stone and the fineness modulus of the sand. Due to WPSA restrictions, the total binder quantity needed to be minimized for a constant water-binder ratio of 0.5. With reference to the C&CI method a larger aggregate size would reduce the total binder quantity necessary for a given mix. Therefore, a 19mm stone size was adopted. Table 2 shows the final mix design for all experimental concrete mix.

Table 2. Final mix design per cubic meter of concrete.

Mix	WPSA [%]	OPC [%]	WPSA [kg/m³]	Water [kg/m³]	Coarse Aggregate [kg/m³]	Sand [kg/m³]
1	0.00	100.00	0.00	205	1017	833
2	5.00	95.00	20.50	205	1017	848
3	10.00	90.00	41.00	205	1017	863
4	20.00	80.00	82.00	205	1017	895
5	30.00	70.00	123.00	205	1017	929
6	*50.00*	50.00	205.00	205	1017	1003

Mixing, Casting and Curing

The making and curing of test specimens was done in accordance with SANS 5661-3:2006. Molding, compaction, and curing were done in accordance with this code. The apparatus used includes moulds (for 100x100x100 cubes) which provide test specimens with the dimension tolerances complying with SANS 5860 and were made of a non-absorbent material (plastic) that is not easily attacked by concrete They were covered with plastic sheets to avoid loss of moisture for 24 hours, before being demoulded and then submerged in portable water until testing.

Compressive Strength Test

The testing of the hardened concrete was done in accordance with SANS 5863:2006 after curing the specimens for 3, 7 and 28 days. The highest load applied, as well as the appearance of the specimen and any unusual characteristics in the type of failure, were all documented. The results are shown in in Table 4 in the next section.

Results and Discussion

Workability of Fresh Mixes: The control mix, with 100% OPC, recorded a slump of 40mm without the addition of a superplasticizer. The benchmark of 40mm was set for the remaining mixes. Therefore, the criterion for a mix to be accepted was a slump between 40mm and 65mm. A significant reduction in workability was observed as the amount of WPSA increased. This may be due to WPSA having a significantly larger specific surface area compared to OPC. All mixes containing WPSA were stiffer and less workable. Enough superplasticizer was therefore added to

Advances in Cement and Concrete   Materials Research Forum LLC
Materials Research Proceedings 51 (2025) 78-84   https://doi.org/10.21741/9781644903537-9

each mix to achieve an acceptable slump value. A larger amount of superplasticizer was required as the percentage of WPSA increased.

Compressive Strength Results: The compressive test results are shown in Fig.3 below. It reveals that at 28 days, the mix containing 5% WPSA attained the highest strength values. This was closely followed by the control specimen but the strength values of the mixtures containing 10%, 20 and 30% were all lower than that of the control mix but significantly higher than the minimum structural strength of 25 MPa. The observed trend is due to the progressive dilution of the active components in the Control mix as the percentage replacement with waste sludge ash increased.

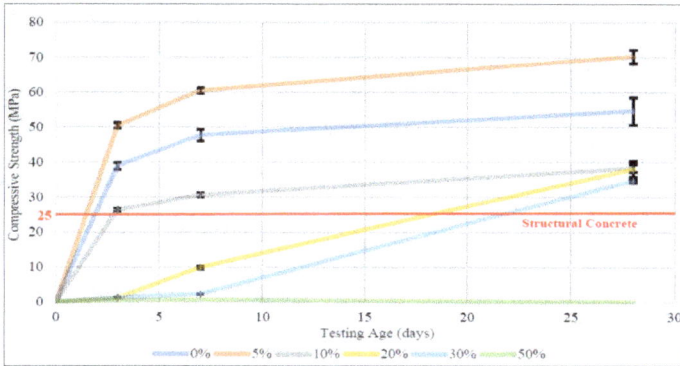

Fig.3. Compressive Strength Development for Varying Portland Cement Replacement Percentages by WPSA.

The specimens incorporating 50% WPSA replacement did not develop any appreciable compressive strength value over the period of 28 days. This is due to the greater dilution of the active mineral components that would have reacted to produce the calcium hydroxide that would normally react with the Silica and Alumina contained in the pozzolan, WPSA, to produce more hydration products, which are then deposited in the concrete to create a more densified microstructure.

Conclusion
The following conclusions can be drawn from the results discussed above:

- Waste paper sludge ash is a viable cement replacement material and can therefore be used to achieve a more sustainable concrete production.
- The study shows that 5% replacement produced a 28% increase in compressive strength compared to concrete made with only OPC.
- 10% replacement produced lower strength than that of OPC but a higher strength than the minimum strength required for structural application.
- 20 % and 30% replacement produced low early strength but achieved a compressive strength higher than the minimum structural strength at 28 days.
- 50% WPSA replacement did not develop any appreciable compressive strength value over the period of 28 days.
- Circular Economy can be achieved in the construction industry by the appropriate use of WPSA in the manufacture of concrete.

References

[1] H.Wu, B. Huang, X. Shu, J. Yin, Utilization of solid wastes/byproducts from paper mills in Controlled Low Strength Material (CLSM), Construction and Building Materials, Vol.118, (2016), 155-163. https://doi.org/10.1016/j.conbuildmat.2016.05.005

[2] M.F. Azrizal, M. N. Noorsuhada, M. F. P. M. Latif, M. F. Arshad, H. Sulaiman, J. Physic: Conference. Series. 1349 012087, (2019). https://doi:10.1088/1742-6596/1349/1/012087

[3] M. I. M. M. B. W. R. A. Ahmad, Journal of Engineering (IOSRJEN), Vol. 3(11), pp. 6-15. 2013. http://doi:10.9790/3021-03113061

[4] S. Nwaubani, Viability of Waste Paper Sludge Ash in Concrete Production, in ARSSS, Int. Conf. London, Edn: 845, SI No 01, 1-5, ISBN:978-93-90150-32-8.

[5] N.Kakora, Partial Substitution of Ordinary Portland Cement with Paper Sludge Waste Ashes in Concrete. Master of Engineering Thesis, University of the Witwatersrand, Johannesburg, South Africa, (2019).

[6] A. Balwaik and S P Raut, Utilization of Waste Paper Pulp by Partial Replacement of Cement in Concrete, International Journal of Engineering Research and Applications (IJERA), Vol. 1, Issue 2, 300-309, ISSN: 2248 – 9622, (2011).

[7] J. Bai, A. Chaipanich, J.M. Kinuthia, M.O'Farrell, B.B. Sabir, S. Wild, M.H. Lewis. Compressive strength and hydration of wastepaper sludge ash–ground granulated blastfurnace slag blended pastes, Cement and Concrete Research 33 (2003), 1189 – 1202. https://doi.org/10.1016/S0008-8846(03)00042-5

[8] M.Sathish, C. R. Babu, A. Sathiya, Performance of Concrete Using Paper Sludge Ash and M-Sand. Irish Interdisciplinary Journal of Science & Research (IIJSR) Vol.5, Iss.3, Pages 26-32, July-September (2021). https://ssrn.com/abstract=3898740

[9] E.Mozaffari, J.Kinuthia, J.Bai, S.Wild, An investigation into the strength development of Wastepaper Sludge Ash blended with Ground Granulated Blastfurnace Slag, Cement and Concrete Research 39(10), 942-949, https://doi.org/10.1016/j.cemconres.2009.07.001

[10] J. Pera, A. Amrouz, Development of highly reactive metakaolin from paper sludge. Advanced Cement Based Materials. 7 (1), 49-56. (1998). https://doi.org/10.1016/S1065-7355(97)00016-3.

[11] M. Mavroulidou, B. Feruku, G. Boulouki, Properties of structural concrete with high-strength cement mixes containing waste paper sludge ash. J Mater Cycles Waste Manag 24(9), 1317–1332, (2022). DOI:10.1007/s10163-022-01402-z

[12] B. A. U. Marvroulidou, 13th International Conference of Environmental Science and Technology. Athens,Greece, London South Bank University, (2013). edium=researchbib https://cest2013.gnest.org

[13] V. Corinaldesi, G. Fava, M.L.Ruello Paper mill sludge ash as supplementary cementitious material, Sustainable Construction Materials and Technologies, Ancona, Italy, (2010). http://dx.doi.org/10.1061/(ASCE)MT.1943-5533.000021,8

[14] M.Frías, R.Garcia, R. Vigil, S.Ferreiro, Calcination of art paper sludge waste for use as a supplementary cement material. 189-193. (2008). https://www.scribd.com/document/461297417/Frias-et-al-2008

[15] P. Jeongjun, G.Hong. Strength Characteristics of Controlled Low-Strength Materials with Waste paper Sludge Ash (WPSA) for Prevention of Sewage Pipe Damage, Materials 13(19), 4238. 23 Sep. (2020). https://doi.org/10.3390/ma13194238.

Advances in Cement and Concrete
Materials Research Proceedings 51 (2025) 85-92

Materials Research Forum LLC
https://doi.org/10.21741/9781644903537-10

An investigation into recycled crushed glass waste as partial replacement for fine aggregate in a self-compacting high-strength concrete

Abubakar Sheriff AJAYE[1,a*], Anthony MUOKA[1,b], Akeem Gbenga AMUDA[1,c], Abdulganiyu SANUSI[1,d]

[1]Civil Engineering Department, Nile University of Nigeria, Abuja Nigeria

[a]asheriff28@gmail.com, [b]anthony.muoka@nileuniversity.edu.ng, [c]akeem.amuda@nileuniversity.edu.ng, [d]sanusiabdulganiyu@nileuniversity.edu.ng

Keywords: Crushed-Glass, Fine-Aggregate, Slump-Flow, V-Funnel-Flow, L-Box

Abstract: The study aimed to investigate the partial replacement of fine aggregate with Crushed Glass Waste (CGW) in Self-Compacting High-Strength Concrete (SCHSC). A total of 84 concrete specimens were produced, replacing Natural Fine Aggregate (NFA) with 5%, 10%, 15%, 20%, 25%, and 30%. The research followed the guidelines set forth by the European Federation of National Trade Associations (EFNARC) for testing fresh and hardened properties of SCHSC. Dry mixture of cement, well graded coarse aggregate of size 20 mm maximum, fine aggregate of size 0.125 mm and crushed glass waste particles passing sieve sizes 4.75 mm to 0.075 mm were added 2.53 kg of clean water with 0.05 kg of superplasticizer within few minutes. After a uniform and homogeneous mixtures of SCHSC were achieved, Slump flow test, V-funnel flow test and L-box test were carried to determine the flowability of fresh concrete. The results indicated that substituting (CGW) as Fine Aggregate (CGWFA) for NFA had no significant impact on the initial slump, with all specimens exceeding the minimum slump criteria of 650 mm. Additionally, the EFNARC requirement for passing ability was met with the results of 7.2 s to 11.6 s for V-funnel flowtime and 0.92 mm to 0.99 mm for L-box. Conclusively, replacement of crushed glass above 20% led to an increased slump flow time at the range of 5.1 s to 5.2 s, indicating a slight decrease in workability. The V-funnel flow times and L-box (3 re-bars) tests satisfied the EFNARC requirements.

Introduction

The relentless march of industrialization and urban expansion has ushered in an era of unparalleled technological advancements, enhancing the quality of life for humankind. However, this progress has come at a substantial cost, manifested in the prodigious generation of non-biodegradable solid waste, posing formidable environmental challenges on a global scale [1]. Improper solid waste management practices, such as indiscriminate dumping and the proliferation of open landfills, have exacerbated the depletion of precious land resources and accelerated environmental degradation [2]. The construction industry, a voracious consumer of natural resources, has recognized the imperative to explore sustainable alternatives to mitigate the depletion of finite natural aggregates employed in concrete production [3].

Waste glass, a non-biodegradable material, has emerged as a viable substitute for natural aggregates in concrete production. The recycling and reuse of waste glass not only address the environmental concerns associated with its indiscriminate disposal but also offer economic benefits by reducing the energy consumption and costs associated with mining and processing natural aggregates [4]. The incorporation of waste glass in concrete production has the potential to enhance the thermal insulation properties of buildings, thereby contributing to energy efficiency and cost savings [5].

Advances in Cement and Concrete

Materials Research Proceedings 51 (2025) 85-92

Materials Research Forum LLC

https://doi.org/10.21741/9781644903537-10

Extensive research endeavours have been dedicated to investigating the utilization of waste glass as a partial replacement for natural aggregates in concrete production, with promising results reported [6]. However, most of these studies have focused on the use of waste glass as a substitute for coarse aggregates, while its potential as a replacement for fine aggregates in self-compacting high-strength concrete (SCHSC) remains relatively unexplored.

Structure of Self-Compacting High-Strength Concrete

SCHSC is a specialized type of concrete that can flow and consolidate under its own weight, without the need for vibration or compaction, while maintaining high compressive strength [7]. The unique properties of SCHSC, such as excellent workability, high durability, and superior mechanical performance, render it an attractive choice for various construction applications, including complex architectural structures and heavily reinforced elements. The ability of SCHSC to flow and self-consolidate without external compaction eliminates the potential for honeycombing, segregation, and other defects commonly associated with conventionally vibrated concrete, thereby enhancing the quality and durability of the finished product. The study was carried-out to investigate the feasibility of partially replacing natural fine aggregates with crushed glass waste (CGW) in the production of SCHSC. The research endeavours to evaluate the fresh and hardened properties of SCHSC containing varying proportions of CGW as a fine aggregate substitute [8]. The findings of this study are expected to contribute to the existing body of knowledge on sustainable concrete manufacturing practices and promote the effective utilization of waste glass as a valuable resource in the construction industry, thereby mitigating the environmental impact of indiscriminate waste disposal and reducing the depletion of finite natural resources.

Table 1: Chemical Composition of Glass Waste [Author, 2024]

Composition	Fe_2O_3	SiO_2	Al_2O_3	MgO	SO_3	TiO_2	MnO	CaO	K_2O	CuO	SrO	Nb_2O_5	Sb_2O_3	CeO_2	BaO
%	0.23	58.8	2.27	2.97	0.29	0.22	0.04	6.57	0.21	0.05	0.45	0.11	0.64	0.03	1.00

Materials

The present study investigates the potential of incorporating crushed glass waste (CGW) as a partial replacement for   natural fine aggregates in the production of self-compacting high-strength concrete (SCHSC). The materials utilized and their respective proportions, as well as the experimental procedures and testing methodologies employed, are comprehensively described in this section.

Cementitious Material

The cementitious binder employed throughout this research was Dangote Ordinary Portland Cement, conforming to the specifications outlined in BS EN 197-1. This cement type is widely used in the construction industry and ensures consistent quality and performance characteristics.

Aggregates

The coarse aggregates used in this study were normal-weight aggregates of granite type obtained from crushed stone quarry, complying with the durability requirements set forth in EN 12620 and EN 206-1. The maximum permissible size of the coarse aggregates was 20 mm, as larger aggregate sizes may adversely affect the flow and passing ability of SCHSC. The fine aggregate used for the research was river sharp sand passing sieve sizes 4.75 to 0.15 accordance with ASTM C136-06.

Glass Waste

Waste glass materials of varying thicknesses were sourced from a finished glass cutting industry located in Utako, Abuja Municipal Area Council. The collected glass waste underwent thorough cleaning to remove dirt and impurities. Subsequently, the glass waste was mechanically crushed

Advances in Cement and Concrete
Materials Research Proceedings 51 (2025) 85-92

Materials Research Forum LLC
https://doi.org/10.21741/9781644903537-10

and sieved to obtain the desired fine aggregate size, suitable for use as a partial replacement for natural fine aggregates in the SCHSC mixes. Personal protective equipment such as eye glass, safety footwear, thick clothing cover, head protective cover, hand gloves were strictly put on the crushing operation to prevent injuries to persons within the area.

Chemical Admixture
In accordance with EN 943, Part 2, Tables 1 and 2, Hydroplast-200D, a high-range water-reducing and retarding admixture, was incorporated to enhance the workability, impermeability, and durability of the SCHSC mixtures. Hydroplast-200D is specifically designed to reduce the water demand while improving the mechanical and physical properties of concrete, making it particularly suitable for applications in hot weather conditions.

Mixing Water
Potable water from the public mains supply was used for mixing the concrete.

Mix Proportioning and Batching
The SCHSC mixtures were batched by weight in the laboratory based on the predetermined combination of proportions. The mixing process involved three distinct stages. Initially, the powder ingredients such as cement and aggregates were dry mixed for a few minutes to ensure homogeneity. Next, half of the mixing water, containing the complete dosage of superplasticizer, was added and mixed for a few minutes. The components were mixed for few minutes to achieve a uniform and homogeneous SCHSC mixture.

Slump flow test, V-funnel flow test and L-box test were carried to determine the flowability of fresh concrete, before SCHSC mixtures were carefully placed into oiled moulds of size 150 mm by 150 mm without any mechanical or manual compaction. The specimens were left undisturbed for 24 hours before demoulding and subsequently cured in a water tank until the desired age for testing the hardened properties.

Table 2: Self-Compacting Concrete- Crushed Waste Glass (SCC-CWG Mix Design) [Author, 2024]

Mix ID	SP (kg)	CWG replacement (%)	W/C ratio	CWG (kg)	Water (kg)	Cement (kg)	FA (kg)	CA (kg)
SCC-CWG-0	0.05	0	0.48	0	2.53	5.21	10.42	10.93
SCC-CWG-5	0.05	5	0.48	0.52	2.53	5.21	9.90	10.93
SCC-CWG-10	0.05	10	0.48	1.04	2.53	5.21	9.38	10.93
SCC-CWG-15	0.05	15	0.48	1.56	2.53	5.21	8.86	10.93
SCC-CWG-20	0.05	20	0.48	2.08	2.53	5.21	8.34	10.93
SCC-CWG-25	0.05	25	0.48	2.61	2.53	5.21	7.82	10.93
SCC-CWG-30	0.05	30	0.48	3.13	2.53	5.21	7.30	10.93

Fresh Concrete Properties
To qualify as a self-compacting concrete, the fresh SCHSC mixture must exhibit the following essential properties: filling ability, passing ability, and segregation resistance. These properties were evaluated through a series of standardized tests, as outlined below.

Advances in Cement and Concrete
Materials Research Proceedings 51 (2025) 85-92

Materials Research Forum LLC
https://doi.org/10.21741/9781644903537-10

Testing Procedures for fresh mixture of SCHSC

Slump Flow Test

Some portion of freshly mixed concrete was poured into a metallic cone with its base resting on a flat metal plate and concrete filled to its top level within some few minutes. The cone was then raised up carefully to a height that allowed the content of the concrete to fall out under gravity while counting the time for fall and spread on the flat plate respectively. The total spread on the plate was measured with a ruler, which represent the slump flow in mm and the corresponding flow time measured with a stopwatch, and both recorded.

V-funnel Test

Some other portion of freshly mixed concrete was poured into a metallic V-funnel resting on its flow collector bucket while bottom gate closed to retain concrete content and concrete filled to its top level within some few minutes. The gate was carefully opened to allow the content of the concrete to pass out under gravity while counting the time it will take for concrete content to pass into the bucket. The total passing time was measured with a stopwatch and recorded.

L-box Test

The separator gate between the vertical and horizontal section of the L-box apparatus was closed and freshly mixed concrete poured into the vertical section until it filled to the required level. The gate was carefully removed to allow the concrete content to pass through the 3-rebars into the horizontal section of the box. The ratio of the concrete content collected by the horizontal section and that of the vertical section was determined in each test.

Results and Discussion

This presents the results of all the tests that were conducted in this research. The tests focus on physical properties on slump flow diameter, V-funnel flow and L-box.

Table 3: Results of Fresh State Properties of SCC Mixtures with different CWG Partial Replacement for Sand [Author, 2024]

Type of test	MIX ID							
	SCC CWG-0	SCC CWG-5	SCC CWG-10	SCC CWG-15	SCC CWG-20	SCC CWG-25	SCC CWG-30	EFNARC Range
Slump flow time(sec)	2.9	3.6	4.5	4.6	5.1	5.15	5.2	2 – 5
Slump flow diameter (mm)	662	668	670	682	693	710	730	650 – 800
V-funnel flow time (sec)	7.2	7.6	8.1	8.3	9.4	10.2	11.6	6 – 12
L-box (h2/h1)(3 rebars)	0.92	0.88	0.82	0.85	0.89	0.91	0.99	0.8 – 1.0

The results obtained from the slump flow test, as illustrated in Fig. 3.1, revealed that the slump flow times were within the range of 2.9 to 5.6 seconds. According to the guidelines set forth by the European Federation of National Associations Representing for Concrete (EFNARC), the suitable slump flow time for self-compacting concrete (SCC) is typically between 2 and 5 seconds. The slump flow times for the SCC-CWG-0 to SCC-CWG-15 mixes were found to be within the acceptable limits prescribed by EFNARC, indicating satisfactory workability. However, the SCC-CWG-20 to SCC-CWG-30 mixes exhibited slump flow times exceeding the recommended range, suggesting a slight decrease in workability. The reason may be because of increase in volume of crushed glass content.

Advances in Cement and Concrete
Materials Research Proceedings 51 (2025) 85-92

Materials Research Forum LLC
https://doi.org/10.21741/9781644903537-10

Specifically, the control mix (SCC-CWG-0) exhibited a slump flow time of 2.9 seconds, while the SCC-CWG-5, SCC-CWG-10, and SCC-CWG-15 mixes recorded slump flow times of 3.6, 4.5, and 4.6 seconds, respectively. These values are indicative of good workability, as they fall within the EFNARC guidelines. In contrast, the SCC-CWG-20, SCC-CWG-25, and SCC-CWG-30 mixes displayed slump flow times of 5.1, 5.15, and 5.2 seconds, respectively, which slightly exceeded the recommended range, suggesting a marginal reduction in workability. Further analysis of the slump flow diameters, as depicted in Fig. 3.1, revealed that all the mixes exhibited slump flow diameters within the range of 650 to 750 mm in accordance to EFNARC standard. According to EFNARC, acceptable slump flow diameters for SCC lie within the region of 650 to 800 mm. Consequently, all the mixes demonstrated satisfactory slump flow diameters, conforming to the established guidelines.

Fig. 3.1: Picture showing Slump Flow Test conducted

The V-funnel flow time test, as illustrated in Fig. 3.2, was employed to assess the viscosity and filling ability of the fresh concrete mixes. The EFNARC guidelines stipulate a V-funnel flow time range of 6 to 12 seconds for SCC. Notably, all the tested mix samples satisfied this criterion during the fresh state evaluation. The control mix exhibited a V-funnel flow time of 7.2 seconds, with a steady increase observed for the subsequent mixes, culminating in a maximum flow time of 11.6 seconds for the SCC-CWG-30 mix. The V-funnel flow times ranged from 7.2 to 11.6 seconds, aligning with the recommended flow time according to EFNARC guidelines.

Fig. 3.2: Picture showing V-Funnel Flow Test conducted.

The L-box test, designed to evaluate the passing ability of SCC, was conducted with three reinforcing bars obstructing the flow path. As presented in Table 3, all the mixes exhibited blocking ratios ranging from 0.88 to 0.99. EFNARC specifies a minimum permissible value of 0.8

Advances in Cement and Concrete
Materials Research Proceedings 51 (2025) 85-92

Materials Research Forum LLC
https://doi.org/10.21741/9781644903537-10

for the blocking ratio, indicating that all the mixes in this study classified as Self-Compacting Concrete (SCC), demonstrating satisfactory passing ability.

Fig. 3.3: Picture showing L-Box Test apparatus of the mixed concrete flow.

Conclusions and Recommendations

Conclusions

Based on the comprehensive experimental investigation into the utilization of glass waste as a replacement for fine aggregates in self-compacting high-strength concrete (SCHSC), the following conclusions can be drawn:

i) The replacement of crushed glass waste as fine aggregate above 20% led to an increased slump flow time at the range of 5.1 s to 5.2 s, indicating a slight decrease in workability.

ii) The V-funnel flow times for all the samples tested satisfied the EFNARC requirements of 7.2 s to 11.6 s

iii) L-box (3 re-bars) tests showed the blocking values of 0.92 to 0.99 satisfying the EFNARC requirements

Recommendations

Based on the findings of this study, the following recommendations are proposed for future research endeavours:

i) Further investigations should be undertaken to explore the potential of alternative surface treatments or modifications to the CGW particles, aiming to enhance their interaction with the cementitious matrix and improve the workability and flow characteristics of SCHSC mixes incorporating higher proportions of CGW as a fine aggregate replacement.

ii) Comprehensive life-cycle assessments and cost-benefit analyses should be conducted to evaluate the environmental and economic implications of incorporating CGW in SCHSC production, considering factors such as energy consumption, greenhouse gas emissions, and the potential for large-scale implementation in the construction industry.

iii) Long-term durability studies should be carried out to assess the performance of SCHSC mixes incorporating CGW under various environmental conditions, including exposure to aggressive agents, freeze-thaw cycles, and other relevant factors that may impact the service life of concrete structures.

References

[1] C. Sun, Q. Chen, , J. Xiao, & W. Liu, Utilization of waste concrete recycling materials in self-compacting concrete, Resources, Conservation and Recycling, 161(2020). https://doi.org/10.1016/j.resconrec.2020.104930

Advances in Cement and Concrete
Materials Research Proceedings 51 (2025) 85-92

Materials Research Forum LLC
https://doi.org/10.21741/9781644903537-10

[2] S. Needhidasan, B. Ramesh, &, S. Joshua Richard Prabu, Experimental study on use of E-waste plastics as coarse aggregate in concrete with manufactured sand, Materials Today: Proceedings, 22 (2020), 715–721. https://doi.org/10.1016/j.matpr.2019.10.006

[3] C. Rodseth, P. Notten, &, H. von Blottnitz, A revised approach for estimating informally disposed domestic waste in rural versus urban South Africa and implications for waste management, South African Journal of Science, (2020),116(1–2). https://doi.org/10.17159/sajs.2020/5635

[4] N. Gupta, R. Siddique, & R. Belarbi, Sustainable and Greener Self-Compacting Concrete incorporating Industrial By-Products: A Review, Journal of Cleaner Production, (2021),284, 124803. https://doi.org/10.1016/j.jclepro.2020.124803

[5] J. Arrieta` Baldovino, J. de, R. L. dos Santos Izzo, É. R. da Silva, &, J. Lundgren Rose, Sustainable Use of Recycled-Glass Powder in Soil Stabilization, Journal of Materials in Civil Engineering, (2020),32(5), 1–15. https://doi.org/10.1061/(asce) mt.1943-5533.0003081

[6] N. Tamanna, R. Tuladhar & N. Sivakugan, Performance of recycled waste glass sand as partial replacement of sand in concrete, Construction and Building Materials, (2020), 239. https://doi.org/10.1016/j.conbuildmat.2019.117804

[7] O. M. Olofinnade, J. M. Ndambuki, A. N. Ede, &, D. O. Olukanni, Effect of substitution of crushed waste glass as partial replacement for natural fine and coarse aggregate in concrete. Materials Science Forum, (2016), 866, 58–62. https://doi.org/10.4028/www.scientific.net/MSF.866.58

[8] A. Gupta, N. Gupta, A. Shukla, R. Goyal, & S. Kumar, Utilization of recycled aggregate, plastic, glass waste and coconut shells in concrete - A review. IOP Conference Series: Materials Science and Engineering, (2020), 804(1). https://doi.org/10.1088/1757-899X/804/1/012034

[9] V. Tanwar, K. Bisht, K. I. S. Ahmed Kabeer, & P. V. Ramana, Experimental investigation of mechanical properties and resistance to acid and sulphate attack of GGBS based concrete mixes with beverage glass waste as fine aggregate, Journal of Building Engineering, 41(2020), 102372. https://doi.org/10.1016/j.jobe.2021.102372

[10] M. N. N. Khan, A. K. Saha, & P. K. Sarker, Reuse of waste glass as a supplementary binder and aggregate for sustainable cement-based construction materials: A review. In Journal of Building Engineering, 28(2020). Elsevier Ltd. https://doi.org/10.1016/j.jobe.2019.101052

[11] W. A. Prasetyo, E. S. Sunarsih, T. L. A. Sucipto, & K. Rahmawati, Enhancing Tensile Strength and Porosity of Self Compacting Concrete (SCC) with Glass Waste Powder. IOP Conference Series: Earth and Environmental Science, 1808(1) (2021). https://doi.org/10.1088/1742-6596/1808/1/012012

[12] J. D. Redondo-Mosquera, D. Sánchez-Angarita, M. Redondo-Pérez, J. C. Gómez-Espitia, & J. Abellán-García, Development of high-volume recycled glass ultra-high-performance concrete with high C3A cement. Case Studies in Construction Materials, 18(2023). https://doi.org/10.1016/j.cscm.2023.e01906

[13] D. Sa, S. M., Su, & A. Ba, Journal of Building Engineering Cleaner prodution of concrete by using industrial by-products as fine aggregate: A sustainable solution to excessive river sand mining. 42(2021)., 1–17. https://doi.org/10.1016/j.jobe.2021.102415

[14] N. Singh, P. Kumar, & P. Goyal, Reviewing the behaviour of high volume fly ash based self compacting concrete. Journal of Building Engineering, 26(2019). 100882. https://doi.org/10.1016/j.jobe.2019.100882

Advances in Cement and Concrete

Materials Research Forum LLC

Materials Research Proceedings 51 (2025) 85-92

https://doi.org/10.21741/9781644903537-10

[15] O. M. Smirnova, I. M. P. de Navascués, V. R. Mikhailevskii, O. I. Kolosov, & N. S. Skolota, Sound-absorbing composites with rubber crumb from used tires. Applied Sciences (Switzerland), 11(2021). https://doi.org/10.3390/app11167347

[16] H. Wei, A. Zhou, T. Liu, D. Zou, & H. Jian, Dynamic and environmental performance of eco-friendly ultra-high performance concrete containing waste cathode ray tube glass as a substitution of river sand. Resources, Conservation and Recycling, 162(2020). https://doi.org/10.1016/j.resconrec.2020.105021

Advances in Cement and Concrete
Materials Research Proceedings 51 (2025) 93-102

Materials Research Forum LLC
https://doi.org/10.21741/9781644903537-11

Curbing the menace of reinforced concrete buildings collapse: A conceptual design and re-design approach for reversing the ugly trends

Timothy Oyewole ALAO[1,a*], James OLAYEMI[2,b] and Aliyu ABDUL[1,c]

[1]Department of Building, Federal University of Technology, Minna. Nigeria

[2]Department of Civil Engineering, Federal University of Technology, Minna. Nigeria

[a]timothy.alao@futminna.edu.ng, [b]olayemi.james@futminna.edu.ng, [c]aliyuabdul79@yahoo.com

Keywords: Conceptual Design, Conceptual Re-Design, Moment Re-Distribution, Load Path, Structural Efficiency

Abstract. Building collapse not arising primarily from the use of poor-quality materials has always been a worrisome concern to both Builders and Structural Engineers. This occurs, despite strict adherence to the fundamentals of structural mechanics employed in structural analysis and design. Case studies of total building collapse and avoidance of near collapse were discussed in the context of providing explanations for causes of collapse and measures employed to avert collapse of reinforced concrete buildings. Measures such as moment re-distribution, re-directing load paths, enhancing greater flexural rigidity, eliminating design flaws and construction sequence/methodology were discussed. Designing such a safe and efficient structural form using this conceptual design and re-design approaches would inherently satisfy both ultimate and serviceability requirements thus avoiding a total or near collapse of reinforced concrete buildings.

Introduction

Building collapse phenomenon

The commonly reported building collapse scenarios have always been attributed to poor quality materials usage on building construction projects [1,2,3]. However, building collapse can also arise from accidents caused by construction plants and equipment, which are often rare reported occurrence and, in such scenarios, sanctions, liabilities and remediations have been meted [4]. In all of these cases, a one-time visual inspection can reveal the causes and blames can be allotted to the regulatory agencies, consulting/supervising teams and the construction team. In contrast, building collapse that has been a worrisome concern particularly to construction professionals, are those scenarios where unexplainable causes are inherent at first sight. Such scenarios defy adherence to basic fundamentals of structural mechanics commonly employed in structural analysis and design [5,6].

Procedures for structural analysis and design

The procedure is guided in the form of design codes of practice and are spelt out in various code of practices which varied within countries such as BS 8110-Part-I and BS EN 1992 [7,8], which are improvements over the Code of Practice CP 114 and CP 110 encompassing the new design philosophy of the limit state principles. This process is guided by basic fundamentals of structural mechanics of materials. Several structural theoretical methods are employed such as the moment distribution method, the flexibility matrix method, the stiffness matrix method, the slope deflection and the clapeyrons (three-moments) methods for studying the nature, type and magnitude of internal stresses including deformation characteristics which reveals and empowers the designers for producing designs that are efficient and safe [9]. It is also practicable to develop mathematical derivations that mimics natures' mechanics to solve a number of structural design concepts. In this

Advances in Cement and Concrete   Materials Research Forum LLC
Materials Research Proceedings 51 (2025) 93-102   https://doi.org/10.21741/9781644903537-11

manner, innovative structural design concepts can be modelled and built. An example of this approach is the adaptation of the Fibonacci sequence which provides inspiration to optimal structural efficiency as demonstrated in the design of the Chinese World Trade Centre [10]. This structural concept mimics the bamboo's unique structural characteristics. It was demonstrated in this concept that the Bamboos diaphragm elements over the entire height are predictable mathematically [10]. The inspiration was curled from the ingenious nature and response of bamboos especially when subjected to Tsunamis to resist lateral forces.

Improving the strength of structural elements using new materials and technological process.
Enhancing the performance of structural elements can be achieved using new materials. The use of modifying agents or additives can enhance strength and durability [11]. Erstwhile, a popular practice has been methods ranging from the use of composite materials with a view to increasing the flexural rigidity of the structural member, increasing percentage of reinforcing steels, increasing the overall dimension and use of prestressed concrete. A good example of increasing the flexural rigidity is in the use of reinforced concrete beam encasing structural steel I-sections which are often limited because of the overall cost implications [11]. Today, concrete is available in the form of self-compacting or self-consolidating concrete (SCC) including the use of super plasticizers and other chemical additives which can enhance greatly, workability, the compressive tensile and flexural strengths with reduced shrinkage [6,9,11,12]. This process enables concrete structures that can now be designed to prescribed level of load bearing and durability. Concrete produced which can now be measured in terms of strength, workability, compatibility, dimensional stability and resilience are now referred to as high-performance concrete (HPC).

Notably, the compressive strength of concrete has erstwhile increased from the highest of $55N/mm^2$ up to $105N/mm^2$ between a threshold of 20 years [11] due to these new developments. Because of earlier limitations on the achievable compressive strength levels till the end of 60's, concrete framed building with the highest number of floors, Pirelli Building in Italy was 127m high. There was a further innovation in the subsequent years. However, in the 90's, the Telekom Malaysian Towers was 310m high and now, the Burj Khalifa Building in Dubai, the United Arab Emirates is 800m tall. This is attributable to the increase in the static efficiency of concrete which can be depicted by Equation (1). The relationship between the compressive strength and unit weight of concrete is now as much as 4.7 times with an accompanying reduced cost which is marginally 3.0 times [11].

$$h_{sc} = \frac{f_c}{P_c} \qquad (1)$$

Where h_{sc} is defined as static efficiency of concrete, f_c is the compressive strength of concrete and P_c is unit weight of concrete.

Improving the strength of structural elements using conceptual design approach
Architectural conceptual design process is also conceivable which are capable of enhancing performance of building fabrics and cost-in-use which can considerably enhance strength and reduce deterioration of structural elements and building fabrics [5]. A conceptual design process primarily refers to an intuitive and knowledge-based reasoning for allocating and maximizing space for functionality, aesthetics and efficiency of an architectural or structural layout. A structural conceptual design process permits an overall development of adequate resistance to strong wind loads, avoids undesirable stress distribution, thereby ensuring robustness of the building. It can also be explored in redirecting load paths aimed at overall optimal structural efficiency while avoiding concepts that creates maintenance concerns [5,6,12,13,14].

Focusing on the overall structural implications by synthesizing the structural system allows alternative structural layout to be evaluated from a multiple conflicting architectural design criterion. While this process can be more time consuming, the task of re-analyzing the structural

Advances in Cement and Concrete   Materials Research Forum LLC
Materials Research Proceedings 51 (2025) 93-102   https://doi.org/10.21741/9781644903537-11

frames can yield a more robust designs that are more efficient and also safe. However, numerous design Softwares which include SAP 2000, STAAD.Pro and Autodesk ROBOT Structural Analysis Professional are available to make this task seamless thus, reducing time consuming tasks of analyzing the structural frames.

The principle of a conceptual design approach should not be considered as a creative approach but as an intuitive reasoning process to create a structure that is not only functional but safe and in addition to this procedure should be at a reduced cost. Some practice procedures and Expert-based geometric modeling and parametric techniques have also been developed and employed to make decisions about the topology and geometry of the concept that is being created, with 3-D models for visualization and manipulation [13,14,15,16]. Authorities such as the United States Federal Highways Authority [15] has developed policies aimed at producing standard practices in the form of guidelines for ascertaining the overall implications of conceptual design solutions.

Balancing theoretical analysis and design
The process of balancing a theoretical analysis and design is aimed primarily, to satisfy the limit state principles of ultimate and serviceability satisfying both strength, deflection and stability [7,8]. This process is often dominated by the constraint of cost and thus, this limit offers designers to devise options that balances this constraint of cost. A balanced design requires an in-depth theoretical knowledge meeting both structural and constructability criteria by utilizing sound theories of structural mechanics and materials.

The use of commercialized/specialized application Softwares has made the task of theoretical structural analysis and design a less tedious and painful task and therefore, the designers are empowered to make judgements in a somewhat painless effort. In addition to computer packages that utilizes basic procedures of structural analysis and design, mathematical derivations that uses natures' mechanics can also be developed and used for structural analysis and also to ascertain the reliability of the design solutions [10,11]. These methods can also be an organically inspired solutions using genetic algorithms process that mimics the evolution of natural reproduction and selection. These methods are nonetheless possible but with the development of mathematical formulations describing the form often referred to as objective functions. The solution procedure is further simplified with the development of constraints formulations and boundary conditions in order to arrive at a feasible solution. This methodology, of course can lead to an array of solutions that the designer is able to make a choice of the 'best' solution.

Methodology

Load paths and re-directing load paths
Load paths can be described as the route within the structural frame along which the load 'flows' through [12]. Primarily, the loads from the roof and floors are transferred unto the beams, and then unto the columns which are then further transferred unto the foundation. Conversely, in a plane truss, the load path flows through the struts and ties unto the supports to foundations. Similarly, the foundation loads are transferred unto the earth. An analysis of the structural system reveals the type and distribution of stresses including the magnitude of displacements which could be horizontal, vertical, rotation or torsion in x, y or z axis. Identifying an efficient load path that minimizes the magnitude of these stresses would imply an efficient design which may not necessarily be a costlier option nor occupying more spaces.

These efficient ways of enhancing the carrying capacity of these members which may be either in flexure, tension, torsion or compression can be carried out through creative or an intuitive reasoning process. Many of the computerized analytical methods such as the global stiffness matrix method can make this task less cumbersome [9]. The design process ensures that all design criteria are met complying with both the ultimate, serviceability and stability requirements. Figures 1(a) and (b) shows an example of an idealized load path for the Pavilion, Raleigh, NC. A method

Advances in Cement and Concrete   Materials Research Forum LLC
Materials Research Proceedings 51 (2025) 93-102   https://doi.org/10.21741/9781644903537-11

such as creating a continuous structural frame, over a number of supports reduces substantially, span moments through the increase of support moments.

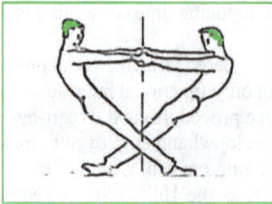

The J. S Dorton Arena Pavilion, Raleigh, NC

(a)   The mechanism   (b) The Bethlehem Steel structure

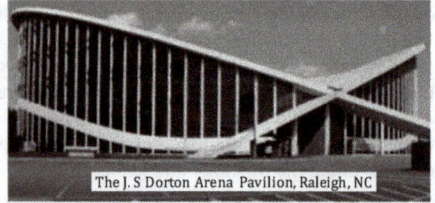

Figure 1: Pavilion, Raleigh Building, (Bethlehem Steel), NC: Courtesy of Fraser, (1981)

Moment re-distribution

In the elastic analysis of structural frames, concrete/reinforced concrete when compared to structural steel does not behave elastically near ultimate loads and for this reason, reinforced concrete frame can only be guaranteed for low stress levels [7,8]. In recognition of this, results of elastic analysis should further be evaluated in order to make the serviceability requirements to be within a tolerable threshold, possibly with values well below the specified limits. The code has recognized this phenomenon that, as the section nears the ultimate moment of resistance, plastic deformation occurs and a re-distribution of the estimated elastic moments can be employed to enhance the structural performance as depicted in Equations (2) and (3), [17]. By considering a span A – B for a uniformly loaded continuous beam, a support moment can be reduced by reducing an excessively large support moment value in Equation (2), and the corresponding shear force can therefore be re-calculated in Equation (3).

$$M_{BA_i} = \left(V_{AB_i} - \frac{w_i L_i}{2} \right) L_i + M_{AB_i} \tag{2}$$

$$V_{AB_i} = \sqrt{(M_{max_i} - M_{AB_i}) 2 w_i} \tag{3}$$

for all supports A_i *and* B_i; *maximum span moment* M_{max_i}; *Current iteration of the shear force* V_i; *span* "L_i" *and udl* "w_i"

The tolerable limit for this re-distribution is 30 percent representing a moment of resistance not greater than 70 percent at the cross section. The neutral axis depth is not to exceed the limit as described in Equation (4).

$$x \leq (0.6 - \beta_{red}) d \tag{4}$$

As a precaution, for frames greater than 4-storeys, this moment reduction should not be greater than 10% particularly for frames in order to avoid lateral instability in the frame. The method is applicable to indeterminate structural frame. In this case, once a beam reaches its ultimate moment of resistance, any further stresses must be taken up partly by the adjacent part of the structure. This moment re-distribution is aimed at maintaining the static equilibrium of the structure. It is important to note that code provisions do not permit redistribution in column moments, [17].

Enhancing greater flexural rigidity

The process of elastic structural analysis and design start with member sizing. The magnitude of internal stresses is dependent among other factors on sizes, shape, form or orientation, length and type of materials used [17]. The flexural rigidity of a structural member can be enhanced by balancing the member properties. The expression in Equation (5) consists of three primary properties of the structure namely, The Young's modulus of elasticity "E" which is essentially a

Advances in Cement and Concrete   Materials Research Forum LLC
Materials Research Proceedings 51 (2025) 93-102   https://doi.org/10.21741/9781644903537-11

material property, the moment of inertia of the member "I" representing the cross-sectional or shape property and the overall length of the member.

$$Flexural\ rigidity = \frac{E_i I_i}{L_i}$$ (5)

$$for\ all\ member\ "i"$$

Secondary considerations such as end fixity, pre-strains and temperature conditions are also to be considered. However, designs that do not conform to mathematically proven solutions or lacking basic principles of structural mechanics are never acceptable. These procedures are basically to avoid structures that do not comply with codes to achieve prescribed safety levels. Engineered structures therefore must exhibit proven mechanics, construction technique, durability and sustainability [11,14, 17]. Where necessary, there may be a need to query a design output to satisfy some basic requirements in order to improve the design output to achieve a functional, stable and a durable structure.

Eliminating design flaws
The process of structural analysis and design are influenced by a number of factors including the use of alternative material or structural form. Change in material specification or use of composite material could sometimes be an un-economic option. In recognition of the requirement for a balanced design, option to change a structural form such as re-designing a section of the sub-frame and exploring an alternative structural layout should be explored.

However, collaborations between design teams to produce functional design solutions is encouraged so as to reduce conflict with architectural design concepts. Code requirements should not be a substitute for intuitive reasoning to produce efficient and safe structures [14,15]. Avoiding undesirable stress distributions and sway can ensure robustness of the building [5,6,10]. Walls are slender members [18] and therefore an exceptionally long and tall walls remains unstable including gable ends of roofs and must be designed to satisfy both strength and stability requirements. A code of practice therefore should not be a substitute for care and vigilance.

Construction sequence/methodology
The sequence of construction is an operational procedure and is well embedded in the practical knowledge acquired during training and practice. This process is usually well outlined and is a required contract document called construction programme representing a step-by-step activity known as tasks and sub-tasks. Nowadays, an evolving concept called Building Information Modelling (BIM), a digital visual representation of the construction process can also be used which allow a digital view and re-view of the overall process. During the construction phase, the structure can be acted upon by strong wind loads or lateral forces which may be foreseeable or unforeseeable. Adequate timing for removal or re-introduction of temporary supports to mitigate against collapse or lateral instability are desirable [6]

Discussion of Results

Scenario 1: Re-design and re-directing load path case study
The example shown in Figures 2(a) - (d) representing the plan and elevation of a storey building with the initial concept where there are four columns in a hall which will obviously obstruct view while the upper floor is to accommodate a residential use. The re-design involves removing the four interior columns resulting in a design output that violates all design criteria. The concept as shown in Figures 2(b) and (d) represents an iterative design output where all design criteria were satisfied. The process in Figure 3 shows the iterative history with the positioning of the columns showing different load path patterns. Column C_1 produced an un-satisfactory design with exceptionally large span positive moment. Column $C2_{L1}$ produced a design with the negative

Advances in Cement and Concrete   Materials Research Forum LLC
Materials Research Proceedings 51 (2025) 93-102   https://doi.org/10.21741/9781644903537-11

moment not reducing to zero showing that it exerts a negative moment on the foundation. However, in column $C2_{L2}$, the negative support moment has reduced to zero and the iteration was terminated because planning law/set back conditions with a combined footing introduced for columns C_1 and $C2_{L2}$.

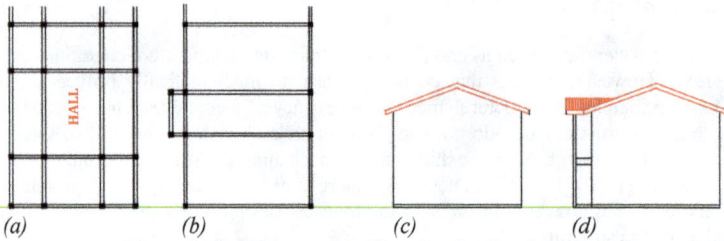

(a)   (b)   (c)   (d)

Figure 2: Re-design and re-directing load path example

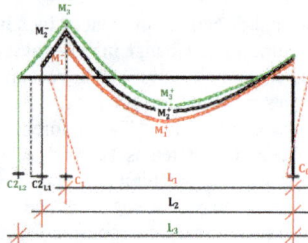

Figure 3: Iterative history of re-design example

The values of the span bending moment can be expressed as shown in Equation (6). Similarly, the maximum span moment can be estimated and occurs at the point where the value of the shear force is zero by equating the value of the expression in Equation (7) to zero. The effects of other loading systems can also be accounted for, in the equations.

$$M_{SPAN_i} = V_{SPAN_i}x - \frac{Wx_i^2}{2} - M_{AB_i} \qquad (6)$$

$$S_{SPAN_i} = V_{SPAN_i} - W_{x_i} \qquad (7)$$

Scenario 2: Moment re-distribution example

The Figure 4 shown is a sub-frame example. The frame has five members, with six joints and three joint restraints are as shown. The design output showed a large negative moment at the interior support labelled M_{R1}^- while the span moment MR_1^+ was small. In accordance with the BS code, when design criteria cannot be met (such as percentage reinforcement, limiting deflection), the interior support moments can be reduced while increasing the span moments as described in Equations (2) and (3). However, the limiting percentage reduction is required not to exceed thirty percent. Where the storey height exceeds three, the limiting percentage reduction is not to exceed ten percent as shown in Equation (3). The iterative history is depicted as shown. In this manner, collapse can be averted using this iterative design procedure.

Advances in Cement and Concrete
Materials Research Proceedings 51 (2025) 93-102

Materials Research Forum LLC
https://doi.org/10.21741/9781644903537-11

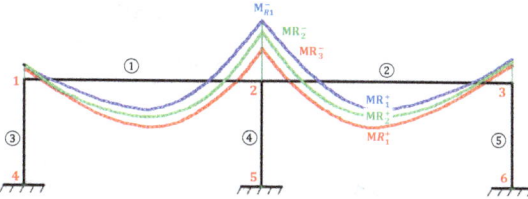

Figure 4: Iterative history of moment re-distribution example

Scenario 3: Eliminating design flaws

The scenario represents a collapsed building. Inspection showed a complete collapse without crushing of the beams. A sway of the entire building occurred, despite that the evaluated strength of the reinforced concrete beam showed a sufficient concrete compressive strength and sufficient number of tensile and compressive reinforcing bars. The plan of the initial frame is shown in Figure 5(a) while the re-built layout is shown in Figure 5(b). The mechanism of the collapse is shown in Figures 6(a) and (b).

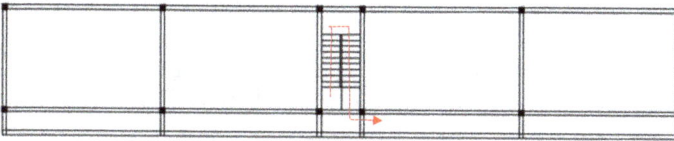

Figure 5(a): The plan of initial frame layout

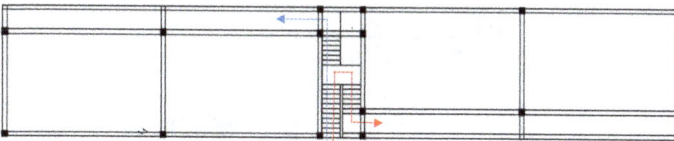

Figure 5(b): The plan of new frame layout

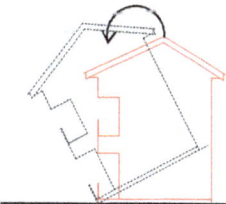

Figure 6(a): The idealized failure mechanism mechanism by sway

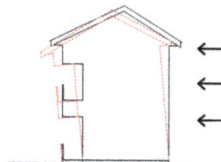

Figure 6(b): The idealized failure by tilting

Scenario 4: Avoiding collapse through ensuring lateral stability

In Figures 7(a) and (b), the point labelled as ①, the gable end collapsed by tilting inwards. This is in realization of the slenderness of the wall. This occurred in a hydraform laterite-cement interlocking brickwall. The collapsed brick did not break into fragments indicating that the brick strength is strong enough but lateral instability is responsible for the collapse. The Figure 7(b) shows a re-built wall with vertical reinforced concrete pier introduced at the beam level. Another alternative solution would be to introduce brick pier wall as shown in the point labelled ②. Avoiding lateral instability in walls are essential for avoiding collapse of walls, fences and gable ends of residential buildings.

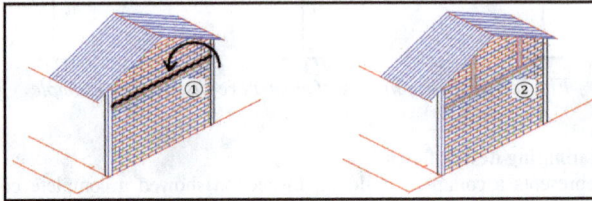

Figure 7(a): Collapse of gable
end of vertical wall

Figure 7(b): Remedial solution for the
gable end

It is also worthy to note that the stability of walls can be enhanced by ensuring the structural members are tied to ensure a rigid frame. This can be achieved with the use of lintels and beams at window levels and at suitable heights to ensure walls do not collapse by lateral instability.

Scenario 5: Avoiding collapse using construction sequence/methodology
The scenario shows a factory building divided into a number of bays as shown in Figure 8. The roof gutter is constructed with reinforced concrete. The walls are carried by reinforced concrete columns and the steel roof truss were connected to the walls with bolted connections.

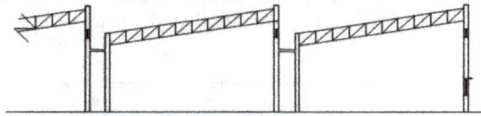

Figure 8: The factory cross-sectional view

Immediately the formworks were removed from the soffits and sides of the roof gutter, there was a lateral movement back and forth as shown in Figure 9(a) which results into a near collapse. Once this scenario was noticed, the solution in Figure 9(b) was used by building temporary sandcrete walls in between the column spaces to stabilize the roof gutter before the roof truss would be installed. In this manner, the roof gutter was stabilized. Similarly, tall walls, otherwise refered to as slender walls are vulnerable and remains unstable and vertical piers built of sandcrete blockwalls or stabilization with reinforced column piers with beams at heights greater than 2.0m are desirable to mitigate against the slender elements in order to obtain a stable structure.

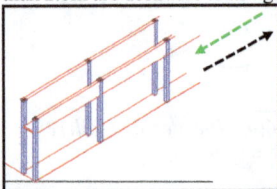

Figure 9(a): Isometric view of factory R.C
roof gutter to stabilize the R.C roof gutter

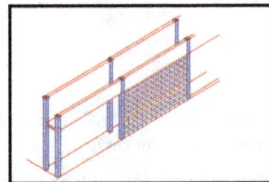

Figure 9(b): Alternating temporary sandcrete
blockwall to prevent lateral movement

Advances in Cement and Concrete   Materials Research Forum LLC
Materials Research Proceedings 51 (2025) 93-102   https://doi.org/10.21741/9781644903537-11

Scenario 6: Enhancing greater flexural rigidity

The frame shown in Figure 10 was analyzed and the resulting bending moment diagram is shown in Figure 11(a). However, the support moments were exceptionally high and the design process showed that limiting requirements were not satisfied while the span moment was small in magnitude when compared with the high support moments. The frame's span moment was designed as a simply supported beam as shown in Figure 11(b), and providing sufficient tensile reinforcement to satisfy both the ultimate limit and serviceability states requirements. Where the architectural concept will not be impaired, doubling the beam in the point labelled ① in Figure 10 could also be an option in order to increase the flexural rigidity of the frame.

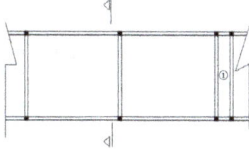

Figure 10: Plan of the frame   Figure 11(a): Initial bending   Figure 11(b): Bending
                                moment   moment of the re-
                                                                 designed frame

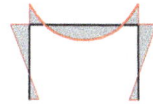

Conclusion

Collapse of reinforced concrete buildings not arising primarily from poor- or low-quality material can occur on construction sites and should be avoided with care, vigilance and intuitive reasoning. The scenarios discussed and represented in schematic diagrams has shown that re-directing load paths, moment re-distribution, enhancing greater flexural rigidity and vigilance/care during construction stage can reduce the ugly phenomenon. This process, obviously would enhance the creative thinking of Structural Engineers and Builders.

References

[1] S. O. Odeyemi, Z. T. Giwa, and R. Abdulwahab. Building Collapse in Nigeria (2009-2019), Causes and Remedies - A Review. Unilorin Science and Engineering Periodicals (USEP): Journal of Research Information in Civil Engineering (RICE). 1(2019) 123-135

[2] D. Obodoh, B. Amade, C. Obodoh, and C. Igwe. Assessment of the effects of Building collapse risks on the stakeholders in the Nigerian built environment. Nigerian Journal of Technology, Faculty of Engineering, University of Nigeria, Nsukka. 38(2019) 822-831. https://doi.org/10.4314/njt.v38i4.2

[3] A. Ede, A. I. Akpabot, S. O. Oyebisi, and F. Gambo. The trend of collapse of buildings in concrete materials in Lagos State, Nigeria (2013-2019). ResearchGate 655(2021). https://doi.org/10.1088/1755-1315/655/1/012078

[4] D. D. Jambol. Curbing the incidences of Building collapse in Nigeria: Sanctions, Liabilities and Legal Imperatives. The Professional Builder. Journal of the Nigerian Institute of Building. 3(2012)

[5] T. O. Alao and E. B. Ogunbode. Enhancing the Performance of Laterite-cement Brick Buildings: A Conceptual Design and Specifications Writing Approach. Proceedings of the 10th West African Built Environment Research Conference,5-7 August. Accra, Ghana. (2019) 508-521.

[6] T. O. Alao and E. B. Ogunbode. Enhancement of the Efficiency of Building Systems through Conceptual Design. Environmental Technology and Science Journal. Federal University of Technology, Minna. 10(2019) 74-83

Advances in Cement and Concrete
Materials Research Proceedings 51 (2025) 93-102

Materials Research Forum LLC
https://doi.org/10.21741/9781644903537-11

[7] BS 8110: Part 1. Structural Use of Concrete - Part 1: Code of Practice for Design and Construction. The British Standard Institute. (1997)

[8] BS EN 1992: Design of concrete structures Part 1: General Rules and Rules for Building. European Committee for Standardization. (2005)

[9] W. Weaver (Jnr) and J. M. Gere. Matrix Analysis of Framed Structures. Third Edition, Van Nostrand Reinhold Company Inc. New York. (1990) 113-173 https://doi.org/10.1007/978-1-4684-7487-9

[10] M. Sarkisian. Tall Building Design Inspired by Nature. 36th Conference on Our World in Concrete and Structures. Singapore, Cipremier, Singapore. (2011) August 14-16.

[11] F. Mola, E. Mola, and L. M. Pellegrini. Recent Developments in the Conceptual Design of R. C. and P. C. Structures. 36th Conference. on Our World in Concrete and Structures. Cipremier, Singapore. (2011) August 14-16.

[12] D. J. Fraser. Conceptual Design and Preliminary Analysis of Structures. Pitman Publishing Inc. London. (1991) 178-193

[13] D. E. Grierson and S. Khajehpour. Method for Conceptual Design applied to Office Buildings, Journal of Computing in Civil Engineering, ASCE, 16(2002) 83-103. https://doi.org/10.1061/(ASCE)0887-3801(2002)16:2(83)

[14] W. Hsu and B. Liu. Conceptual design: Issues and challenges. Computer-Aided Design. 32(2000) 849-900. https://doi.org/10.1016/S0010-4485(00)00074-9

[15] FHWA. Conceptual Studies and Preliminary Design. United States Federal Highway Authority. (2012).

[16] R. Mora, H. Rivard and C. Bedard. Computer-Aided Conceptual Design of Building Structures: Geometric Modelling for the Synthesis Process. Proceedings of the 1st International Conference on Design Computing and Cognition. MIT, Cambridge, USA. (2004) 37-55. Kluwer Academic Publishers. https://doi.org/10.1007/978-1-4020-2393-4_3

[17] W. H. Mosley, J. H. Bungey, and R. Hulse. Reinforced Concrete Design. 5th Edition. Macmillan Publishers, London. (1997) 48-53.

[18] BS 5628-1. Code of practice for the use of masonry. Structural use of unreinforced masonry. British Standard Institution. (2005)

Advances in Cement and Concrete
Materials Research Proceedings 51 (2025) 103-110

Materials Research Forum LLC
https://doi.org/10.21741/9781644903537-12

The effect of water quality on the performance of bamboo leaves ash blended cement concrete

Kazeem Dele MUSBAU[1,a,*], J.O. ADEOSUN[1,b], A.O. ABEGUNDE[1,c]

[1]Department Of Building, Osun State University, Osogbo, Osun State, Nigeria

[a]kazeem.musbau@uniosun.edu.ng, [b]julianah.adeosun@uniosun.edu.ng,
[c]adetomi.abegunde@uniosun.edu.ng

Keywords: Concrete, Bamboo Leaves Ash, Compressive Strength, Water Absorption, Borehole Water and River Osun Water

Abstract. This study investigates the effect of River Osun Water (ROW) and Borehole Water (BHW) on the performance of concrete made with bamboo Leaves Ash (BLA) blended cement of 5% and 10% replacement. It compares the compressive strength (CS) and percentage water absorption (%WA) of BLA blended cement with Portland limestone cement concrete with a view to exploring the suitability of ROW for concrete production. Mix design was carried out using the British Research Establishment (BRE) design guide at free water content of 200 kg/m^3 and water-cement ratios (w/c) of 0.35 and 0.50. The chemical analysis of the mixing water samples was done at Osun State Water Cooperation, Ede. The CS was determined in accordance with British Standard at expiration of 7, 14, 21, 28, 42 and 60 days while %WA was determined to standard at 28-, 42- and 60-days curing. The data obtained were analyzed using means, percentages and graphical illustration. The results showed that the CS of concrete made with ROW follows the same trend with BHW for all the binders that is the CS increases with ages while the %WA decreases with ages at 0.35 and 0.5 w/c. It was also found that the concrete made with ROW were characterized by a slightly lower CS and higher %WA for all the samples compare to BHW. The study concluded that the heavy metals present in ROW participated in the cement hydration. Therefore, ROW is not suitable for the production of high-performance concrete unless where standard compressive strength and durability are not required.

Introduction

Water is not only meant for domestic purposes; it is also one of the ingredients in the production of concrete. Studies have shown that water plays a major role during the concrete production and hardening stage [1, 2]. Therefore, water is Involved in all the concrete stages such as mixing, compaction and curing stages but the understanding of some people is that water for concrete has nothing to do with sources, colour and quality. Aside the fact that water is an essential concrete ingredient, it's also an initiator in cement chemical reaction known as cement hydration to achieve a target characteristic strength [3]. In order to play this role, water for concrete should not contain any harmful impurities and chemicals like clay, organic particles, oils, soaps, detergents, dissolved salts, acids, alkalis, sugars and lime in solution [4]. After the basic concrete materials (aggregates and cement) are adequately mixed, then the predetermined quantity of water will be added and thereafter, the concrete is formed. The mixture starts to set (hard) due to chemical reaction between Portland cement and water, this is called cement hydration. This process continues until the cement paste and aggregates are linked together. Though, numerous research works have been carried out to understand the interaction of Portland cement with different water such as rain water, sewage water, car wash water, pond water and others. There is scanty research work on the performance of blended cement concrete mixed with River Osun water.

River Osun flows southwards from Ekiti State through the Osun State into the Lagos Lagoon in south-west, Nigeria. The river is one of the major rivers in southwest region of Nigeria. It is a common practice for cities located along the flow of river osun utilizing the river's water for both domestic purpose and construction works. In 2018, suddenly the river began to change colour due to the impact of gold mining within the catchment of Osun River. As a result of this, the investigation by Urban Alert (a civic-tech non-profit organization) revealed that unregulated gold mining activities around the river osun is the root cause for the color change. The mining activities have contaminated the river with heavy metals such as gold, mercury and cadmium which occur naturally in the earth's crust, thereby threatening the river Osun water quality. Aside from heavy metals contamination, the river which was once transparent enough, is now very turbid (cloudy) with a characteristic gold colour. In view of this, the ecological condition of the river water especially around the Ijesha land in Osun State down to the Lagoon meeting point is called for investigation to ascertain it is suitability for concrete production since the concrete performance is a function of water quality.

The elements present in the water for mixing concrete will involve in the chemical reactions and thus affect the setting time, hardening, strength development and concrete durability at large. The general believe is that if water is not suitable for drinking, it is also not good for making concrete. That is, polluted water is not suitable for concrete production [5]. Going by this assertion, it will be so difficult to produce a durable concrete due to scarcity of potable water in many cities of the southwest region. In Nigeria, different sources of water are used for concrete production and these have major effect on the strength of concrete due to different levels of impurities [6]. This study intends to check the effect of ROW on concrete performance. The objectives of this study are to determine the compressive strength and water absorption of BLA blended cement concrete, Portland limestone cement concrete using water sourced from Osun River and Bore Hole inside the Osun State University, Oshogbo Campus. Also, the study compared the concrete binders' performance to determine the suitability of mixing water for concrete production.

Materials and Methods

The study adopted two mixing water samples for the experimental work. The River Osun Water (ROW) which was gotten at Eastern Bypass Road, Ijetu Bridge, Osogbo at Lat. 7.743722°, Long. 4.55615° as shown in fig. 1(a) and the Bore Holes Water (BHW), was taken inside the Osun State University, Oshogbo Campus, Faculty of Engineering, at Lat. 7.7585820°, Long. 4.5999016°. These two water samples were tagged ROW and BHW respectively. The collected water samples were analyzed to determine the physico-chemical compositions at the Osun State Water Cooperation Ede, Osun State, Nigeria. The study used three different binders namely: PLC which is 0% replacement, 5% and 10% replacement with BLA. The binders were used to produce concrete with 0.35 and 0.50 water-to-cement ratio and free water content of 200kg/m3 by using Building Research Establishment (BRE) mix design guide (Teychenne et al., 1997). Batching was done by weight for the concrete ingredients with the help of scale calibrated in kilogram. The ingredients were mixed manually on a neat impermeable surface. The measured blended cement was spread over the appropriate measured fine aggregates (sand) and mixed until a uniform colour observed by visual inspection. The required measured coarse aggregate (granite) was added to the already mixed blended cement-sand, and remix thoroughly before the addition of mixing waters (ROW and BHW). Mixing was assumed to be completed when homogeneous mix is obtained by visual inspection. The workability of the fresh formulated concrete mixes was measured using the slump cone mould. The concrete cubes of 100mm by 100mm by 100mm were made with the two mixing water samples at two w/c (0.35 and 0.5) in three and two replicates for determination of CS and %WA respectively. The concrete cubes specimens were form in a wooden mould as shown in fig 1 (a) and (b). The cubes were kept in a place free from vibration and direct rays of sunlight and demoulded after 24 hours and immediately subjected to curing in water tanks for 7, 14, 21, 28, 42 and 60 days for compressive strength test and 28, 42 and 60 days for water absorption test at

Advances in Cement and Concrete

Materials Research Forum LLC

Materials Research Proceedings 51 (2025) 103-110

https://doi.org/10.21741/9781644903537-12

room temperature. The compressive strength was determined using OKH-2000D compressive testing machine in accordance with [7] as shown in fig 1 (d).

Fig. 1 (a) River Osun Water (b) Wooden Formwork (c) Concrete Cubes Samples and (d) OKH-2000D Compressive Testing Machine

The study was conducted on six different concrete mix samples with two mixing waters and various curing ages as shown in Table 1. The total number of concrete cubes required for the study is hundred and eighty-eight (288), out of which two hundred and sixteen (216) cubes were tested for compressive strength and the value was determined after the expiration of each curing ages while seventy-two (72) cubes were characterised for water absorption property. The water absorption test was also carried out after the concrete specimens have been water cured for 28, 42 and 60 days, in accordance with the test method described in [8]

TABLE 1: *Number of concrete cubes required for the study*

Mix	Binders	W/C	Test carried out	Mixing water	Cubes replicate	Curing age (day)	No. cubes required
1	PLC	0.35	Compressive Strength	BHW &ROW	3	7,14,21,28,42,60	36
			%Water Absorption	BHW &ROW	2	28,42,60	12
2	5%BLA	0.35	Compressive Strength	BHW &ROW	3	7,14,21,28,42,60	36
			%Water Absorption	BHW &ROW	2	28,42,60	12
3	10%BLA	0.35	Compressive Strength	BHW &ROW	3	7,14,21,28,42,60	36
			%Water Absorption	BHW &ROW	2	28,42,60	12
4	PLC	0.50	Compressive Strength	BHW &ROW	3	7,14,21,28,42,60	36
			%Water Absorption	BHW &ROW	2	28,42,60	12
5	5%BLA	0.50	Compressive Strength	BHW &ROW	3	7,14,21,28,42,60	36
			%Water Absorption	BHW &ROW	2	28,42,60	12
6	10%BLA	0.50	Compressive Strength	BHW &ROW	3	7,14,21,28,42,60	36
			%Water Absorption	BHW &ROW	2	28,42,60	12
Total							288

Results and Discussion

In the first stage of the research, both ROW and BHW were evaluated in terms of compliance with appropriate agencies [9]. The result of chemical analysis of the water is given in the Table 2, which shows the comparison with max. allowable value set out by standard and agency [9, 10]. From the Table 2, it is clearly shown that the two mixing water samples used for the research work have their pH value within the acceptable limit specified by [9, 10] and for the other metallic ions shown, they are all regarded as impurities in the mixing water because potable water is expected to contain only two elements; Hydrogen and Oxygen with allowance for negligible amount of other non-deleterious elements [9, 10].

TABLE 2: *The chemical analysis for the two mixing waters samples*

Parameters	River Osun Water	Bore Hole Water	WHO (2008 and 2011) Standard	BS-EN 1008 Max. Value
Appearance	Not clear(brown)	Clear	-	-
Colour (H.U)	100.00	15.00	-	-
pH at laboratory	7.2	7.2	6.5-8.5	≥ 4
Iron(mg/l)	3.70	0.00	0.3	-
Silica(mg/l)	1.25	0.36		-
Nitrate nitrogen (NO3^{2-})(mg/l)	0.273	0.041	50	500
Nitrite nitrogen (NO2^{2-})(mg/l)	0.140	0.025	0.2	500
Copper (mg/l)	2.10	0.22	1	-
Manganese (mg/l)	0.008	0.000	0.2	-
Aluminum (mg/l)	0.19	0.00		-
Fluoride (mg/l)	0.150	0.00		-
Sulphide (mg/l)	0.13	0.00		-
Chromium (mg/l)	0.32	0.04		-
Sulphite (mg/l)	48	0.00		-
Sulphate (mg/l)	59	20	100	2000
Potassium (mg/l)	3.6	2.0		-
Phosphate (mg/l)	91.20	12.2		100
Zinc (mg/l)	0.99	0.18		100

The compressive strength

The results of compressive strength for the six (6) concrete samples at 7, 14, 21, 28, 42 and 60 curing ages were presented in Table 3. It shows that at early curing age like 7, 14 and 21 days, Portland limestone cement concrete has highest compressive strength for the two mixing samples compared with BLA blended cement but at later ages (42 and 60 days) the BLA blended cements gain higher strength for BHW and insignificant low for ROW at both w/c. This could be as a result of delay in pozzolanic reaction with calcium hydroxide released from cement and the presence of heavy metals in ROW which hinders the full reaction of bamboo leave ash. The table also shows that 10% replacement blended cement concrete mixed with ROW has lowest strength values across all the curing age due to unfavorable reaction between the heavy metal present in the ROW and increase in BLA percentage replacement.

Materials Research Forum LLC
https://doi.org/10.21741/9781644903537-12

TABLE 3: *The compressive strength for the two mixing waters*

MIX	Cement Replacement	W/C	Compressive strength in days for BHW [N/mm²]						Compressive strength in days for ROW [N/mm²]					
			7	14	21	28	42	60	7	14	21	28	42	60
1	0%	0.35	17.9	19.3	24.3	28.6	30.5	33.4	16.8	19.0	22.4	25.5	28.3	29.7
2	5%	0.35	16.1	18.7	24.1	27.9	31.3	34.2	15.9	18.5	22.0	23.9	27.5	29.0
3	10%	0.35	15.8	18.5	24.5	28.7	32.1	34.9	15.6	17.9	21.8	23.6	27.1	28.9
4	0	0.50	16.5	18.8	23.9	28.0	29.2	31.9	16.3	18.2	21.9	24.5	27.8	29.2
5	5%	0.50	15.4	18.2	23.5	27.9	29.9	32.7	15.2	17.5	20.3	23.8	27.1	28.9
6	10%	0.50	15.0	17.9	23.8	28.2	30.7	33.1	15.2	16.4	20.0	22.7	26.6	27.7

The compressive strength of the two mixing water samples across the curing for all the mix samples as shown in Fig. 2. It shows that with increasing curing ages, the compressive strength increases and with increasing w/c, the compressive strength decreases for the two mixing waters. In general, the compressive strength value for BHW is within the acceptable standard while the concrete mixed with ROW achieved lower compressive strength for all mixes. This shows that ROW is not suitable for producing high performance concrete. This could be as a result of the presence of heavy metals in the ROW which participated in the cement chemical reaction with the aggregates. This implies that ROW can only be used to make concrete of lesser characteristic strength. This supports the existing study [11], which reveals that concretes made with polluted water are characterized by a slightly lower compressive strength than concretes made with tap water/drinking water

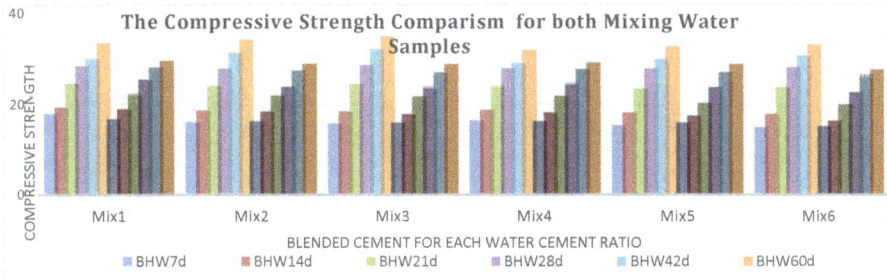

Fig. 2: The compressive strength comparison of each mixing water at different curing ages

The water absorption

The %water absorption of all the three blended cement concrete (0%, 5% and 10% BLA replacement) with w/c (0.35 and 0.5) for the two waters samples across the curing ages was presented in Table 4. It was determined using the formula shown in Eq. 1

$$\text{Percentage water absorption} = \frac{W2-W1}{W1} \times 100 \qquad \text{Eq. [1]}$$

Where W_1 = Dry weight, W_2 = Wet weight, W_3 = W_2-W_1 [Wet weight- Dry weight]
W_4 = W_3/W_1

TABLE 4: *The computation of water absorption test for the two mixing water samples*

Mix	Binder	W/C	Mixing Water Sample	28days				42days				60days			
				W_1	W_2	W_3	W_4	W_1	W_2	W_3	W_4	W_1	W_2	W_3	W_4
1	0%BLA	0.35	BHW	2.00	2.16	0.16	8.00	1.96	2.08	0.12	6.06	1.99	2.08	0.09	4.52
			ROW	2.10	2.28	0.18	8.57	1.97	2.10	0.13	6.60	1.99	2.09	0.10	5.02
2	5%BLA	0.35	BHW	2.03	2.19	0.16	7.88	1.92	2.03	0.11	5.73	1.94	2.02	0.08	4.12
			ROW	1.91	2,07	0.16	8.38	1.92	2.04	0.12	6.25	1.91	2.00	0.09	4.71
3	10%BLA	0.35	BHW	2.09	2.25	0.16	7.65	1.85	1.95	0.10	5.41	1.95	2.03	0.08	4.10
			ROW	2.09	2.26	0.17	8.13	1.80	1.91	0.11	6.11	1.95	2.04	0.09	4.61
4	0%BLA	0.50	BHW	1.98	2.19	0.21	10.61	1.94	2.11	0.17	8.76	1.98	2.08	0.1	5.05
			ROW	1.92	2.13	0.21	10.90	1.99	2.17	0.18	9.05	1.93	2.04	0.11	5.70
5	5%BLA	0.50	BHW	1.97	2.17	0.20	10.15	1.92	2.08	0.16	8.33	1.97	2.06	0.09	4.57
			ROW	1.91	2.11	0.20	10.47	1.95	2.12	0.17	8.72	1.90	2.00	0.10	5.26
6	10%BLA	0.50	BHW	1.98	2.18	0.20	10.10	1.98	2.14	0.16	8.08	1.98	2.07	0.09	4.55
			ROW	2.04	2.25	0.21	10.29	1.99	2.16	0.17	8.54	1.95	2.05	0.10	5.13

The comparison of %WA test at various curing ages for the two mixing water samples was presented in Table 5. The results shows that concrete made with BHW has lower %WA compare to ROW. This indicates that concrete made with ROW absorbed much water for all the concrete samples regardless of the curing ages and w/c. The results also reveals that the replacement of PLC with BLA reduces the water absorption property of concrete due to higher silica content in BLA which involves in the cement chemical hydration.

TABLE 5: *The percentage of water absorption results for both mixing water*

Mix	Blended cement	W/C	Curing ages for BHW			Curing ages for ROW		
			28	42	60	28	42	60
1	PLC	0.35	8.00	6.06	4.52	8.57	6.60	5.02
2	5%BLA	0.35	7.88	5.73	4.12	8.38	6.25	4.71
3	10%BLA	0.35	7.65	5.41	4.10	8.13	6.11	4.61
4	PLC	0.50	10.61	8.76	5.05	10.90	9.05	5.70
5	5%BLA	0.50	10.15	8.33	4.57	10.47	8.72	5.26
6	10%BLA	0.50	10.10	8.08	4.55	10.29	8.54	5.13

The %WA for each mix was shown in Fig. 3. It shows the %water absorption for all the concrete samples at various curing ages for the two mixing waters. It is clearly seen in Fig. 3 that the %water absorption of all the concretes mixes reduces as the curing ages increases. This could be as a result of the pore refinement arising from the filling up of the voids in concrete by the continued

Advances in Cement and Concrete

Materials Research Forum LLC

Materials Research Proceedings 51 (2025) 103-110

https://doi.org/10.21741/9781644903537-12

hydration products formed by hydration reaction of cement with age. This result is in consonance with existing study [12] which observed a decrease in absorption properties of concretes with increase in curing age. However, for the two mixing waters, the percentage water absorption increases with increasing water/cement ratio as a result of decrease in the cement content in the higher w/c ratio. This result supports the existing study [13]. At equal water-binder ratios, the percentage water absorption of concrete with ROW is higher compared to the concrete made with BHW at all curing ages and binders. This shows that at equal w/c, curing age and binder, the percentage water absorption of ROW is higher.

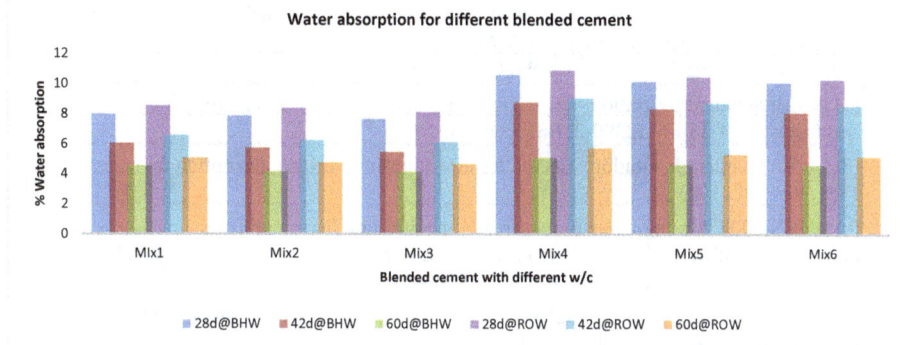

Fig 3: *% water absorption comparison for the two mixing waters*

Conclusions

This research was carried out to investigate the effect of water quality on the blended cement concrete properties such as compressive strength and water absorption. The study analyzed the effect of w/c on BLA blended cement concrete with use of both BHW and ROW. Then, the compressive strength and water absorption test were conducted on concrete made with these water samples. Based on the results of this experimental study, the following conclusions are drawn:

- The quality of water has significant effect on the concrete performance. The heavy metals present in ROW participates in the cement-aggregates chemical reactions and thus affect the strength development and absorption property of the concrete. Thus, concretes made with BHW attains practical standard of compressive strength and absorption property while ROW should not be recommended for concrete production of high standard for now especially in reinforced concrete because of the presence of heavy metals.

- The results of this study constitute the basis for further research that would include concrete durability performance over longer curing period and also the evaluation of the effect of combined different water source on concrete performance because globally, drinking water for domestic use is becoming scarce, hence this type of research gains more importance in order to use the available drinking water for human consumption alone.

References

[1] American Society for Testing and Materials C94/C94M-99, Standard Specification for Ready-Mixed Concrete 1999.

[2] PN-EN 1008, Mixing water for concrete. Specification of sampling, testing and assessment of suitability of mixing water for concrete, including water recovered from concrete production processes, 2004.

[3] K. Wanda, Impact of water quality on concrete mix and hardened concrete parameter. Journal of Civil and Environmental Engineering Reports, (2019). DOI: 10.2478/ceer-2019-0033.

[4] A. Neville, Water and Concrete- A love- hate relationship point view. Journal of Concrete International. (2009), 22: 34-38

[5] H. S. Rao, V. V. Reddy and S. G. Vaishali, Effects of Acidity Present in Water on Strength and Setting Properties of Concrete. Proceedings of 29th Conference on Our World in Concrete and Structures: 25 - 26 August, 2024

[6] O. Ata, Effects of Different Sources of Water on Concrete Strength: A Case Study of Ile-Ife. Civil and environmental research. Vol.6, No.3, (2014), http://www.iiste.org/Journals/index.php/CER/article/

[7] British Standard Institution. Testing harden concrete: Compressive strength of test specimens,2009 BS EN 12390-2, BSI, London.

[8] British Standard Institution. Method for determination of water absorption for concrete. BS 1881: Part 122: 1983

[9] BS EN 1008. Mixing water for concrete - Specification for sampling, testing and assessing the suitability of water, including water recovered from processes in the concrete industry, as mixing water for concrete. (2004) Centre for European Standard.

[10] World Health Organization (2011). Guidelines for drinking-water quality. 4th ed. WHO Press Geneva, Switzerland. (2011) 541p. Available at: www.who.int/water_sanitation_health/publications

[11] P. Woyciechowski, E. Szmigiera and D. Reluga, Impact of recovered water used as mixing water on the characteristics of concrete and concrete mix. Conference Concrete Days, Wisła 2008

[12] B. G. Reddy, R. H. Sudarsana, and I. V. Ramana Reddy, Use of Treated Industrial Waste Water as Mixing Water in Cement Works. Nature, Environment and Pollution Technology Journal, 6 (2007) 595-600

[13] O. Alawode, and O. I. Idowu, Effects of Water-Cement Ratios on the Compressive Strength and Workability of Concrete and Lateritic Concrete Mixes. The Pacific Journal of Science and Technology. 12 (2011) 99- 105.

Advances in Cement and Concrete
Materials Research Proceedings 51 (2025) 111-117

Materials Research Forum LLC
https://doi.org/10.21741/9781644903537-13

Assessment of construction project performance in Nigeria: A case study of Kano State Metropolis

Mujittafa SARIYYU[1,a] *, Nasiru Zakari MUHAMMAD[1,b], Ashiru SANI[1,c], Naziru Ado MUSA[2,d], and Umar Faruk LAWAN[2,e]

[1]Department of Civil Engineering, Faculty of Engineering, Aliko Dangote University of Science and Technology, P.M.B. 3244, Wudil, Kano, Nigeria

[2]Department of Water Resources and Environmental Engineering, Faculty of Engineering, Aliko Dangote University of Science and Technology, Wudil, Kano, Nigeria

[a]mujittafasariyyu@kustwudil.edu.ng, [b]mnasiruzakari@gmail.com,
[c]ashirusani@kustwudil.edu.ng, [d]nmusadam@gmail.com, [e]umarfaruklawan@kustwudil.edu.ng

Keywords: Construction Industries, Construction Sites, Projects and Project Management

Abstract. The elements included in project management start from goals and end with the end product. Undesirable project execution is the biggest issue influencing development industries all over Nigeria and generally creating nations, where endeavors are being made to utilize project execution evaluation to screen and control projects to guarantee favorable results. The objectives are to determine the performance dimension of construction industries within Kano Metropolis and to identify and analyze the factors affecting the performance of construction projects. The data used was through a designed questionnaire. 100 questionnaires were distributed to the project participants of which 70 were returned. The data was analyzed using Simple Descriptive Statistics including Frequency distribution, Percentages, Tables, Bar-chart and Pie-chart, to assess the performance of the construction project. This research indicated that the most important construction stages are the designing, constructing, tendering, briefing, and commissioning stages which possessed 32.9%, 24.3%, 21.4%, 14.3%, and 7.1% respectively. The most important performance dimensions are; quality, time, and cost. The factors affecting project performance are delays, changes in material prices, lack of qualified personnel, poor quality equipment, political reasons, and weather conditions.

Introduction

A project is defined a project as an "Endeavour in which human, material and financial resources are organized, to undertake a unique scope of work, of a given specification, within the constraint of cost and time, to achieve beneficial quantitative and qualitative objectives [1]." A project can be defined as "a temporary endeavor undertaken to create a unique product, service or human and non-human resources pulled together into a temporary organization to achieve a specified objective [2,3].

The Project Management Team includes five stages from the time when the decision is made to implement the construction project until the project becomes a reality [6,7]. These stages are:

i. Briefing Stage: the purpose of this stage is to enable the client to specify project functions and permissible costs so that the architects, engineers, quantity surveyors, and other members of the design team can correctly interpret his wishes and provide cost estimates.

ii. Designing Stage: To complete the project brief and determine the layout, design, method of construction, and estimate costs, to obtain the necessary approvals from the client and authority involved. Also, to prepare the necessary production information including working drawings and specifications, and to complete all arrangements necessary for obtaining tenders.

Advances in Cement and Concrete
Materials Research Proceedings 51 (2025) 111-117

Materials Research Forum LLC
https://doi.org/10.21741/9781644903537-13

iii.   Tendering stage: this is purposely to appoint a contractor or contractors who will undertake the site construction works. Activities are to obtain tenders from contractors for the construction of the building and to award the contract.

iv.   Constructing stages: The purpose of this stage is to construct the structure within the agreed limits of cost and time and to specified quality. The constructing stage consists of some unrelated activities. The failure of one activity can disrupt the entire production schedule.

v.   Commissioning Stage: The purpose of this stage is to ensure that the building (structure) has been completed as specified in the contract documents and that all the facilities work properly. Consider Figure 1 for the Project Management Team.

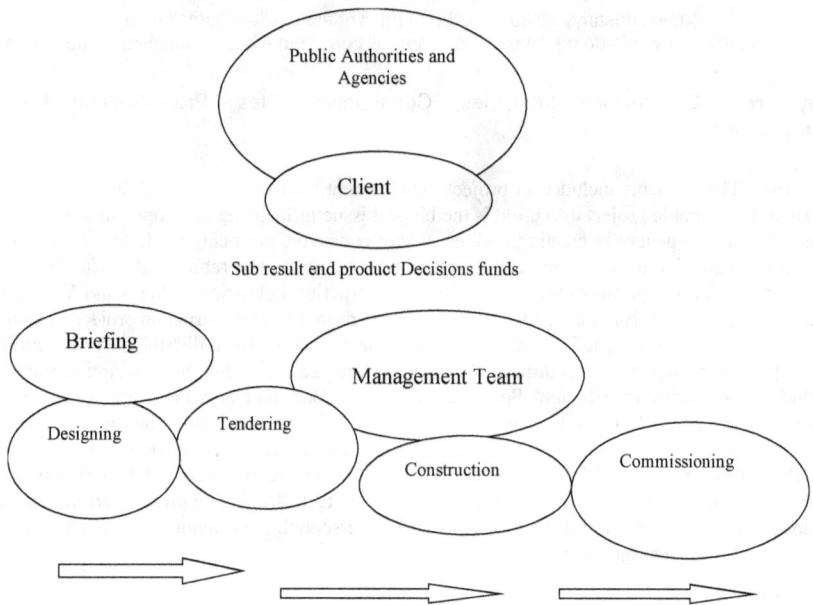

Fig. 1. Project Management [8]

[9] says, management is a practice, not a science, it is not knowledge but performance and it takes place at all levels of the organization. The essence of management is to produce a low maintenance cost. The elements included in project management start from goals and end with the end product as stated by [10,11]. According to [12,13], proper construction management should provide a cycle of activities to achieve the project goals as shown in Figure 2.

Advances in Cement and Concrete
Materials Research Proceedings 51 (2025) 111-117

Materials Research Forum LLC
https://doi.org/10.21741/9781644903537-13

GOALS

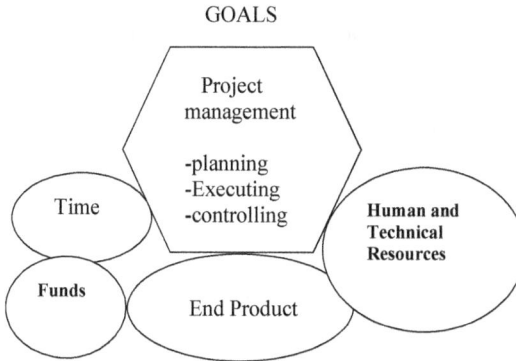

Figure 2: Elements in Project Management [14]

Problem Statement

The undesirable project performance results in various forms such as low productivity, delays, cost overruns, poor quality, and so on. Undesirable project performance is the main problem affecting construction industries everywhere in Nigeria and mostly developing countries, where efforts are being made to use project performance assessment to monitor and control projects to ensure favorable outcomes. Therefore, there is a need to emulate approaches to ensuring improvement in project performance in Nigeria. This assessment provides a wealth of information on several projects which become case histories from which learning, -improvement, and development in project execution can be achieved.

Aim and Objectives

This research aims to assess construction project performance in Nigeria. The following objectives were set:

- To determine the performance dimension of construction industries within Kano Metropolis,
- To identify and analyze the factors affecting the performance of construction projects

Limitation

This research is limited to construction professionals such as engineers, clients, contractors, Architects, quantity surveyors, and builders. Our research work is limited to some selected construction industries within the Kano State metropolis.

Methodology

This research was conducted through a questionnaire designed in such a manner to obtain responses that could be easily analyzed by the use of open-ended questions with suggested answers by the professionals that are capable of attempting such questions (i.e. Engineers, Builders, Quantity Surveyors, Architects, and Contractors). The data was analyzed using Descriptive Statistics including tables, percentages, frequency distribution, Bar-chart, and Pie-chart, to assess construction project performance in Nigeria. 100 questionnaires were distributed of which 70% returned. This research determined whether the ongoing project is succeeding or failing to achieve the objectives for which they are being implemented.

This development is possible because the performance assessment in the form of monitoring and controlling is effective project management.

Advances in Cement and Concrete
Materials Research Proceedings 51 (2025) 111-117

Materials Research Forum LLC
https://doi.org/10.21741/9781644903537-13

Results and Discussion

Figure 3 shows the Profession of the respondents, and Figure 4 displays the performance dimension. Tables 1 and 2 indicate the Project Stages and Factors Affecting the Performance of construction projects respectively.

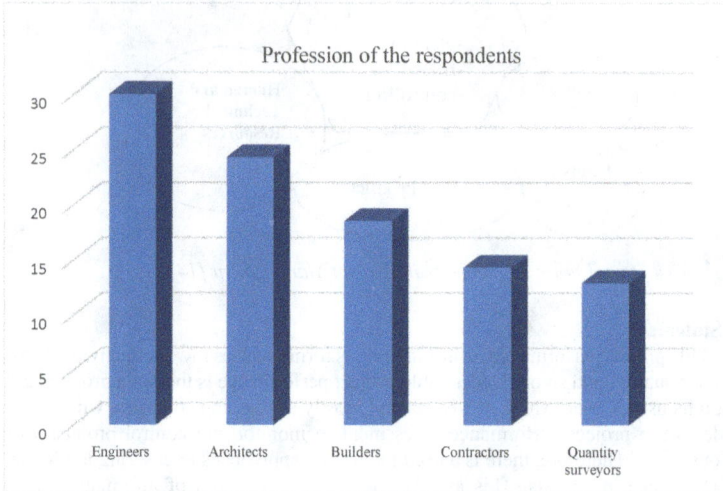

Fig. 3. Profession of the respondents

It was shown that engineers had the highest rank possessed 30%, and Architects had 24.3%. Builders, Contractors, and Quantity Surveyors had 18.5%, 14.3%, and 12.9% respectively. It was shown that engineers are the dominant professionals in gathering information for this research.

Table 1. Project Stages

Stages	Frequency	Percentage	Rank
Designing Stage	23	32.9	1
Tendering Stage	17	24.3	2
Briefing stage	15	21.4	3
Constructing stage	10	14.3	4
Commissioning Stage	5	7.1	5
TOTAL	70	100	

The designing Stage was the first rank with 32.9%, followed by the Tendering Stage which had 24.4%. The briefing stage, constructing stage, and commissioning stage had 21.4%, 14.3%, and 7.1% respectively. These stages identified the importance of the project level from the beginning up to the completion of the project. Our findings indicated that the Designing stage is the most important level and key to consider in the assessment of the project performance in Nigeria (First

Advances in Cement and Concrete   Materials Research Forum LLC
Materials Research Proceedings 51 (2025) 111-117   https://doi.org/10.21741/9781644903537-13

rank), while the least important stage was the Commissioning Stage which ranked 5th position at the project stage.

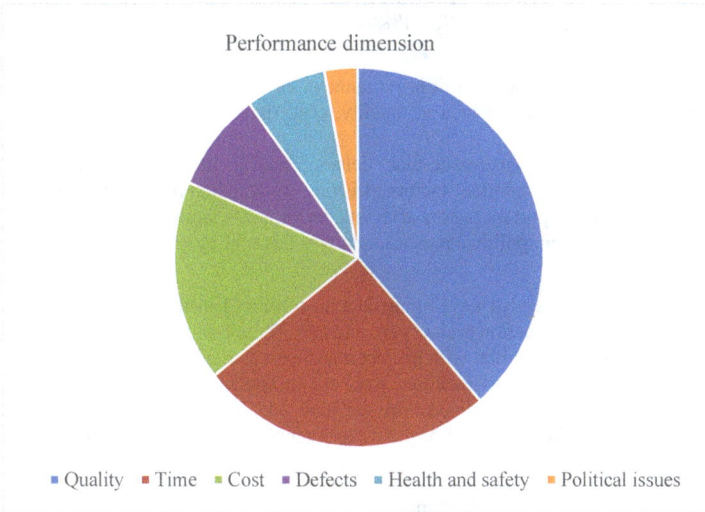

Fig. 4. Performance Dimension

From the results obtained, respondents measured their projects by quality 38.6%, 25.7% measured based on time, and 17.1% used cost dimension, 8.6% measured their projects by their defect, 7.1% measured their projects based on health & safety, and only 2.9% of the respondents measured their project based on political issues. This study indicated that the respondents measured their project performance dimension based on the quality of the project, time of completion, and cost implications due to their highest percentage. Few respondents measured the success or failure of their project by the dimension of defects, health & safety, and political issues.

Table 2. Factors Affecting the Performance of Construction Projects

Responses	Frequency	Percentage	Rank
Delay	19	27.1	1
Changes in materials prices	14	20	2
Incompetent personnel	12	17.1	3
Unavailability of resources	10	14.3	4
Poor quality equipment	8	11.4	5
Political reasons	5	7.1	6
Weather condition	2	2.9	7
TOTAL	70	100	

Delay is the major factor affecting the construction project in the Nigerian construction industry which consists of 27.1%. The changes in material prices were 20%, but Incompetent personnel had 17.1%. The unavailability of resources was 14.3%. Poor quality equipment, Political issues, and weather conditions had 11.4%, 7.1%, and 2.9% respectively.

Conclusions

This research indicated the level of importance of each stage of a construction project. The most important performance dimension is; quality, cost, time, defects, health & safety, and political issues.

There is significant agreement that, delays, changes in material prices, and incompetent personnel are the most important factors affecting construction projects from the perspective of 5 groups, while other important factors affecting performance factors are unavailability of resources, poor quality equipment, political reasons, and weather conditions.

Recommendations

This research recommended developing human resources in the construction industry through proper training in construction project performance. These can update participants' knowledge and assist them to be more familiar with project management techniques and processes.

Clients are encouraged to facilitate payment to contractors to overcome the delay, and all managerial levels should participate in sensitive and important decision-making. Contractors should not increase the number of projects that cannot be performed successfully.

References

[1] E. Adnan and S. A. Sherif Mohamed, Factors affecting the performance of construction projects in the Gaza strip, Journal of Civil Engineering and Management. (2009) 269-280. https://doi.org/10.3846/1392-3730.2009.15.269-280

[2] W. E. Kihoro M., Factors affecting Performance of Projects in the Construction Industry in Kenya, The Strategic Journal of Business & Change Management. (2015) 38-54. https://doi.org/10.61426/sjbcm.v2i2.121

[3] S. D. D., Study of Factors Affecting Performance of Construction Project, International Journal of Science and Research. (2013) 1339-1341.

[4] S. M. Rowlinson, An Analysis of Factors Affecting Project Performance in Industrial Building, Brunel University, Hong Kong, 1988.

[5] B. D.J, Method of Managing different Perfective of Project Success, British Journal of Management, (2005) 111-131.

[6] S. A. C. T. T. Beatham, A Critical Appraisal of their use in Construction Industry, Journal 11, (2004) 93-117. https://doi.org/10.1108/14635770410520320

[7] A. A. J. Brown, Measuring the Effect of Project Managements of Construction Outputs, Journals of Project Management 18, (2000) 327-335. https://doi.org/10.1016/S0263-7863(99)00026-5

[8] B. Z. T. Elyamany A. Isma'il, Performance Evaluating Model for Construction Companies, Journals 133 (8) (2007) 574-581. https://doi.org/10.1061/(ASCE)0733-9364(2007)133:8(574)

[9] P. H. J. &. S. Barrett, "Good practice in briefing: the limit of rationality automation in construction, (1999) 633-642. https://doi.org/10.1016/S0926-5805(98)00108-3

[10] O. Kolawole, A Review of Abandoned Projects Cases in Nigeria, National Association of Building student (2006) 33-40.

Advances in Cement and Concrete
Materials Research Proceedings 51 (2025) 111-117

Materials Research Forum LLC
https://doi.org/10.21741/9781644903537-13

[11] R. Takim, "Performance indicator for successful construction project performance, in Association of Researchers in construction management, UK, 2002.

[12] F. P. Mir, Exploring the value of project management, International Journal of Project Management, (2014) 203-215.

[13] A. Odeyinka and H. Yusuf, The causes and effects of construction delays on the cost of housing project in Nigeria, Journal of Financial Management and Property Construction, (1997) 31-41.

[14] N. R. BELEIU Ioana, Main Factors Influencing Project Success, Interdisciplinary management research, (2010) 59-71.

Advances in Cement and Concrete
Materials Research Proceedings 51 (2025) 118-127

Materials Research Forum LLC
https://doi.org/10.21741/9781644903537-14

Optimization and modeling of fresh state properties of cement-calcined clay-calcium carbide waste pastes using the response surface methodology

A.S. OGUNRO[1,a*], M.A. USMAN[1,b], E.E. IKPONMWOSA[2,c], and R.U. OWOLABI[1,d]

[1]Department of Chemical and Petroleum Engineering, University of Lagos, Akoka, Yaba Lagos101017, Nigeria

[2]Department of Civil and Environmental Engineering, University of Lagos, Akoka, Yaba Lagos101017, Nigeria

[a]steveayodele20012001@yahoo.com, [b]musman@unilag.edu.ng, [c]eikponmwosa@unilag.edu.ng, [d]rowolabi@unilag.edu.ng

Keywords: Modelling, Optimization, RSM, C-S-H Gel, Setting Time, Pozzolanic Reaction

Abstract. A satisfactory model for predicting the setting times has been a challenge, due to the multivariable nature of cement hydration and pozzolanic reactions. This presents a challenge in creating an appropriate model to predict the setting times of pastes. This study presents a model for predicting setting times of pastes formed from partially substituting Ordinary Portland Cement (OPC) with calcined Ifonyintedo clay (CIC) and calcium carbide waste (CCW). Individual and interactive effects of three processing factors on setting times were investigated by a Vicat apparatus using the central composite design (CCD) model of response surface methodology (RSM) for experimental design, modeling, and process optimization. The modeled optimization conditions were CEMII (77.1wt%), CIC (17.3wt%), and CCW (5.64wt%), water-binder (w/b) ratio (0.244) achieving desirability of 1 with the corresponding initial and final setting times (IST, FST) of 118.65mins, 312.43mins against the predicted values of 114.71mins, 300.68mins, respectively. The ANOVA confirms that the model is satisfactory, hence, useable for construction works. Comprehensive analysis revealed that silica precursor underwent polymerization, resulting in the formation of products of reaction that exhibit characteristics of C-S-H gel.

Introduction

The increasing demand for sustainable and environmentally friendly construction materials has led to the development of alternative cement-based systems. One such system is the combination of Ordinary Portland Cement (OPC), calcined clay, and calcium carbide waste (CCW), which has shown promising results [1,2]. However, the fresh state properties of such systems play a crucial role in determining their constructability and overall performance. The optimization of fresh state properties involves the interaction of multiple variables. One-factor-at-a-time (OFAT) approach is impracticable and may not provide an understanding of the interactions between factors. The Response Surface Methodology (RSM) is a statistical technique that helps to develop a polynomial model that describes the relationship between the input variables and the response variable. Several studies have revealed that for proper cement's quality control, its consistency and setting times can provide significant information and allow cement handling [1, 2, 3 4]; cement paste may be influenced by the addition of supplementary cementitious materials (SCMs, pozzolanic and hydraulic) like calcined clay etc., and other admixtures like calcium carbide waste (CCW) for partially replace OPC to improve their performance and lower carbon footprint [2, 4, 5, 6]. Setting times rose at low metakaolin (MK) levels and a drop at 15% replacement; in contrast to pastes of silica fume (SF) [2]. Fly ash (FA) caused an increase in the setting time of concrete while

Advances in Cement and Concrete   Materials Research Forum LLC
Materials Research Proceedings 51 (2025) 118-127   https://doi.org/10.21741/9781644903537-14

retardation depended upon the source of FA [1,3]. [3] reported that consistency, setting times at low replacement levels of PC by MK were similar to that of the PC which rose by about 10 and 15 minutes respectively at higher replacements. 1wt % addition of CCW presented setting times of 78 % and 57 % of control values, but increased in consistency by 14 %; 5 wt% CCW increased consistency but decreased the setting times [7, 8]. RSM has found application in production of concrete mortars, optimization of raw materials mix, calcination temperature, contents of rubberized concrete [9, 10, 11 ,12 ,13, 14, 15] achieving maximum compressive strength at lower cost. Researchers have experimented with combining calcined clay and OPC [15, 16 ,17, 18] through OFAT methods. However, not much work has taken into account blending calcined clay and OPC through statistical design method during cement industry production. In this study the partial substitution of OPC with calcined clay and CCW was optimized using the RSM's central composite design (CCD) model. Using CCW to replace OPC during cement production is an environmentally benign way to dispose of materials that would otherwise damage land and water. The RSM based on CCD is more robust to handle losses of runs or mismeasured responses [10]. Hence, this work aims to optimize and model the fresh state properties of the calcined clay-calcium carbide waste cementitious system using the RSM based on CCD. [18, 21] demonstrated that cal cined Ifonyintedo clay (CIC) was preferred over other clays in the same locality because of its ar chitectural and marketing benefits. The detailed objectives of our investigation were (i) to condu ct experimental studies that resulting in the determination of the consistency, initial and final sett ing times using CIC and CCW in the partial replacement of OPC (ii) to identify the optimal mixt ure proportions; (iii) to develop a mathematical model that describes the relationship between the setting times and the process factors (iv) to detect the structural functionality of pastes.

Materials and Methods

Materials

Portland-limestone-cement (CEM II B-L, 42.5 N)-"3X", and portable water were used as stipulated by [22, 23]; Calcined Ifonyintedo clay (CIC) produced by the Nigerian Building and Road Research Institute (NBRRI), Calcium carbide waste (CCW) collected from dumpsites and kept under normal laboratory conditions. Data on the oxide compositions, as previously considered in an aspect of our study [18], are shown in Table 1. The Brunauer–Emmett–Teller (BET) specific surface values measured via the Single-Point and Multi-Point BET model indicated 1.81-3.47×10^2 m^2/g, pore diameter was 3.00 nm (mesoporous material) in line with values of 10-20m^2/g reported for pure kaolinite [21, 22]. These values show that CEM II is the finest due to different production methods.

Table 1 Characterization of Raw Materials

SAMPLE (%)	SiO$_2$	Al$_2$O$_3$	Fe$_2$O$_3$	CaO	MgO	SO$_3$	K$_2$O	Na$_2$O	Mn$_2$O$_5$	P$_2$O$_5$	TiO$_2$	LOI
(CIC)@6002H	65.42	23.2	2.63	1.91	0.26	0.00	0.1	0	0.11	0.21	2.98	3.28
ASTM: C618-08 CLASS N	(SiO$_2$+Al$_2$O$_3$+Fe$_2$O$_3$) ≥ 70%			(CaO + MgO) ≤5%		4% Max						10% Max
CCW	4.32	1.25	0.27	80.45	0.00	0.06	0.00	0.00	0.02	0.07	0.1	12.46
CEMII	16.14	4.48	2.27	60.38	1.21	1.97	1.51	0.00	0.10	0.45	0.31`	11.35

Methodology

Determination of the Setting Times and Standard Consistency

The Vicat apparatus was used in compliance with [23] and the laboratory experiment was carried out at 25 ± 2 ∘C temperatures and 70 ± 5% humidity. Mixing was carried out in a Kenwood Chef Major KM250 mixer by [23]. The CEMII, CEMII-CIC and CEMII-CIC-CCW were blended, and a total solids content of 320g per admixture was investigated. The paste's CEMII component was partially substituted with ground calcined clay CIC at different cement replacement levels of 20%, 40%, 50%, and 60% in a binary mix. At the same time, holding the composition of CEMII constant, 5wt.% CIC was replaced by CCW in the initial binary mix to form a ternary mix of the two admixtures (CEMII/CIC/CCW). Finally, while holding CIC constant, 5wt.% CEMII was replaced by 5wt.% CCW to form a ternary mix of the admixtures (CEMII/CIC/CCW). The paste without CIC and CCW served as the control.

Design of the Expe riment

Experimental runs for optimization studies using RSM were designed using the Central Composite Design, a subdivision of RSM in the program software Design-Expert 10.0.3.0 (Stat-Ease, Inc., Minneapolis). Process factors were examined at four different levels (Tables 2). The factors were chosen in line with the findings of setting times test conducted in section 2.2.1. CEM II, CIC, CCW, and H_2O were varied to examine setting times (IST and FST) as presented i n Table 2. Table 3 presents the experimental runs (calculated by eqn. 1);

$$N = 2^n + 2n + N_c = 2^4 + (2 \times 4) + 4 = 28 runs \qquad (1)$$

Table 2: Setting Times Factors varied using Central Composite Design

Factors	Units	Coded Variable Levels				
		$-\alpha$	-1	0	1	α
CEM II	g	192	224	256	288	320
CIC	g	48	64	80	96	112
CCW	g	8	16	24	32	40
H_2O	g	70	80	90	100	110

Table 3 The Design Matrix, Experimental and Predicted Setting Times

Run	CEMII (g)	CIC (g)	CCW (g)	Amount of water (g)	Observed Initial setting time (min)	Predicted Initial setting time (min)	Observed Final setting time (min)	Predicted Final setting time (min)
1	224	96	32	80	90.00	90.00	330.00	328.79
2	224	16	32	80	140.00	140.60	280.00	280.54
3	288	16	16	80	85.00	84.60	286.00	286.93
4	256	96	24	90	119.00	120.34	340.00	339.75
5	224	96	16	100	120.00	118.33	270.00	270.85
6	224	56	24	90	115.00	116.78	300.00	307.42
7	224	16	16	100	110.00	111.43	230.00	227.10
8	256	56	24	90	125.00	124.83	306.00	307.12
9	288	16	16	100	90.00	89.89	289.00	291.02
10	256	56	24	80	122.00	123.12	295.00	294.09
11	288	16	32	80	130.00	131.55	270.00	269.96
12	288	96	16	80	102.00	102.50	310.00	311.18
13	288	96	32	80	110.00	108.21	290.00	292.21
14	224	166	16	80	122.00	119.89	259.00	258.02
15	288	96	32	100	130.00	132.00	340.00	341.79
16	224	16	32	100	135.00	134.39	302.00	301.63
17	256	16	24	90	128.00	128.56	307.00	306.75
18	288	56	24	90	115.00	115.12	329.00	321.09
19	256	56	24	100	130.00	130.78	303.00	303.42
20	256	56	24	90	130.00	124.83	305.00	307.12
21	256	56	24	90	125.00	124.83	308.00	307.12
22	224	96	16	80	110.00	110.54	310.00	308.27
23	288	96	16	100	125.00	124.04	310.00	308.77
24	288	16	32	100	140.00	139.10	325.00	326.04
25	256	56	16	90	112.00	114.78	266.00	267.86
26	224	96	32	100	100.00	100.04	345.00	343.38
27	256	56	32	90	130.00	129.12	298.00	295.64
28	256	56	24	90	125.00	124.83	308.00	307.12

The pastes prepared with optimum processing conditions were characterized after setting using Cary-630 Fourier-Transform Infrared Spectroscopy (FTIR) produced by Agilent Technologies, Malaysia, as described in our previous study [18] at the Covenant University, Ota, Ogun State, Nigeria.

Advances in Cement and Concrete   Materials Research Forum LLC
Materials Research Proceedings 51 (2025) 118-127   https://doi.org/10.21741/9781644903537-14

Results and Discussions

Characterization

The CEMII spectra presented a narrow and intense band at 3421 and 3380cm^{-1}(Figure 1) caused by vibration vas O-H in the structure of Si-O [24, 25]. Broadband covering 800–1300 cm^{-1} which has a shoulder at 1013 cm^{-1} was caused the vibration of Si-O antisymmetric stretching from wollastonite of SiO_4^{4-} groups and 872 cm^{-1} shoulder assigned to CO_3^{2-} (δ C-O)[24].The characteristic intense and narrow band at 3462 cm^{-1} is attributable to the vas O-H vibration in the $Ca(OH)_2$ structure; CO_3^{2-} bands at 872 (δ C-O) and 1397 cm^{-1} (associated with carbonation).

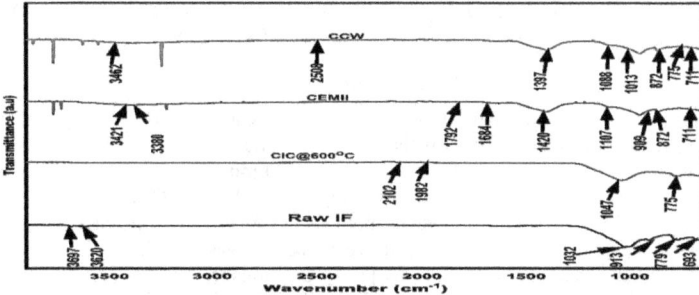

Figure 1 FTIR data for the CCW, CEMII, CIC and raw IF

Kaolinite in raw IF clay (Figure 1) showed two O-H stretching bands at 3697, and 3620 cm^{-1}, while absorption band at 693cm^{-1} was assigned to the Si-O stretching of kaolinite or quartz [16, 18, 24]. The band at 913cm^{-1} is related to the Al-O-H deformation and connected to octahedral and tetrahedral sheets in the kaolinite structure [24]. Calcination at 600 °C resulted in the disappearance of the OH stretching bands and the Al-O-H deformation band at 912 cm^{-1}. This indicates complete dehydroxylation of kaolinite.

Setting Times and Consistency

Standard consistency values of various CEMII-CIC, CEMII-CCW and CEMII-CIC-CCW mixes were found to vary from 29.4 to 36% (Figure 2). These conforms the range recommended for OPC. An increase of CCW fraction in the mix, results in a reduction of the standard consistency value. The specific surface of CIC and CCW are 34 and 66.7 m^2/g respectively and CIC have the plasticity in the order of kaolinites while CCW is non-plastic. Owing to high specific surface area of CIC, a mix containing higher fraction of CIC, requires more amount of CCW solution to lubricate the particle surface and thus it exhibits greater consistency value. Similar results were reported by [1, 4]. Higher fraction of CIC makes the paste plastic, which in turn limits the penetration of the plunger attached to the Vicat apparatus; increasing consistency. Thus, it can be concluded that the standard consistency values of blended CEMII-CIC, CEMII-CCW and CEMII-CIC-CCW pastes are the function of chemical composition, fineness of source materials and the fraction of the replacement materials. The CCW ensures that the Ca/Si ratio is maintained. [3, 7] reported increased setting times with CCW mortar over plain mortar and metakaolin – OPC blends at standard consistency varied from 0.3 to 0.45.

Advances in Cement and Concrete
Materials Research Proceedings 51 (2025) 118-127

Materials Research Forum LLC
https://doi.org/10.21741/9781644903537-14

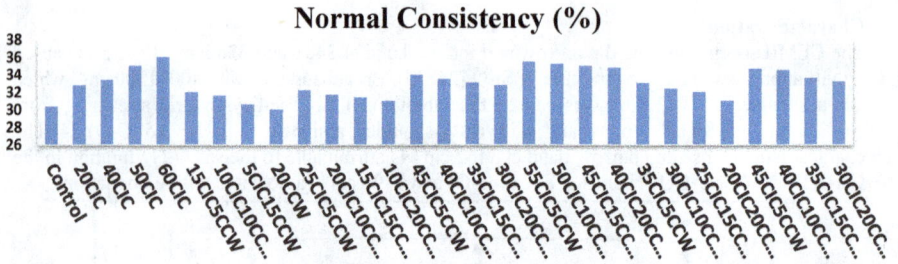

Normal Consistency (%)

Figure 2 The consistencies of the mixes

The initial setting (IS) and final setting (FS) of CEMII-CIC, CEMII-CCW and CEMII-CIC-CCW mixes (Figures 3, 4 5a & 5b) show that IS times of the mixes are found to vary from 90 to 125 min whereas the FS times vary between 270 and 340 min. This is owing to slow chemical reaction of CIC & CCW with time at ordinary temperature. At 5%CIC and 5%CCW the setting times were similar to the CEMII. However, both the IS & FS times increase with increasing CCW&CIC content, although, the increase above 10%CCW was significantly greater than other intervals. A similar remark was also made by [1, 3]. Ca^{2+} ions from CCW happen beside release of Si^{4+} and Al^{3+} ions from the CIC particles. Higher Ca^{2+} ions in CCW supports the pozzolanic reaction besides hydration reaction. These reaction products increase the setting times; this is useful in mass concreting and in hot weather. However, the shift in setting time curve at 15%CIC5%CCW and 25%CIC5%CCW was not in a defined pattern; which is similar to the manner setting time of the CIC–CEMII blends varied with CIC content at 20%CIC replacement of CEMII (Figures 3, 5a & 5b). Similar results were reported by [1, 3, 4].

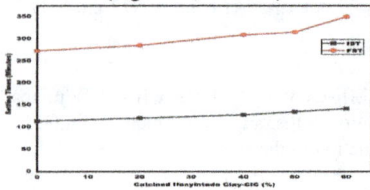

Figure 3 consistency of CIC-CEMII

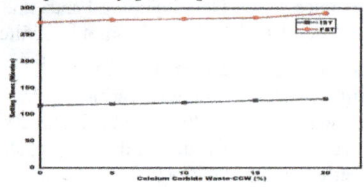

Figure 4 Consistency of CCW-CEMII

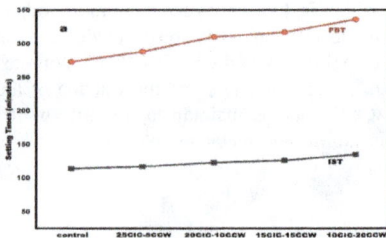

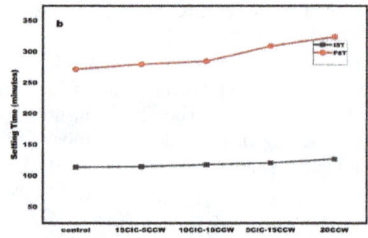

Figure 5a Setting times of CIC-CCW-CEMII pastesFigure 5b Setting times of CIC-CCW-CEMII

Optimization studies

The ANOVA yielded R^2 values of at least 0.9885, meaning that the models can account for at least 98% of the variation in data and that the models did not explain 2% of the total variations. An acceptable model should have an R-squared value of at least 0.75 [20]. Adj R^2 of at least 0.9768; "Pred R-Squared" of at least 0.9505 suggested that the model was very relevant (Table 4 & Figures 6 and 7).

Table 4 The quadratic models with respect to the variables' coded values

Response	Regression Equation	R^2 (%)	Pred. R^2(%)	Adj R^2(%)	Lack of fit
IST	124.83-0.83A-4.11B+7.17C+3.83D+6.81AB+6.56AC+3.44AD-10.31BC+4.06BD+0.56CD-8.88A²-0.38B²-2.88C²+2.12D²	98.88	95.05	97.68	0.73
FST	307.12+683A+16.50B+13.89C+4.67D-6.50AB-9.87AC+8.75AD-0.50BC-1.62BD+13CD+7.13A²+16.13B²-25.37C²-8.37D²	99.08	95.61	98.10	7.25

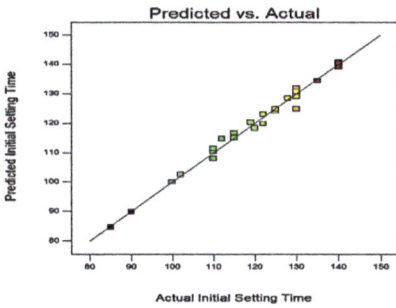

Figure 6: IST experiment vs predicted values   *Figure 7: FST experiment vs predicted values*

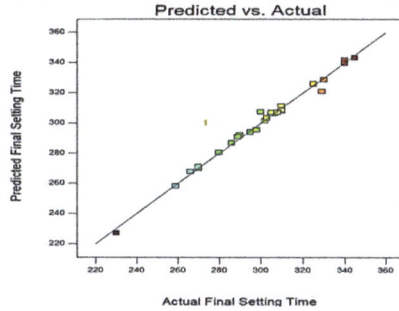

Table 5 Quadratic Model ANOVA for Initial Setting Time

Source	Sum of Squares	df	Mean Square	F Value	p-value Prob > F	
Model	5724.04	14	408.86	82.23	< 0.0001	Significant
A-CEMII	12.50	1	12.50	2.51	0.1368	
B-CIC	304.22	1	304.22	61.19	< 0.0001	
C-CCW	924.50	1	924.50	185.95	< 0.0001	
D-W/B	264.50	1	264.50	53.20	< 0.0001	
AB	742.56	1	742.56	149.35	< 0.0001	
AC	689.06	1	689.06	138.59	< 0.0001	
AD	189.06	1	189.06	38.03	< 0.0001	
BC	1701.56	1	1701.56	342.24	< 0.0001	
BD	264.06	1	264.06	53.11	< 0.0001	
CD	5.06	1	5.06	1.02	0.3314	
A²	203.25	1	203.25	40.88	< 0.0001	
B²	0.37	1	0.37	0.074	0.7903	
C²	21.35	1	21.35	4.29	0.0587	
D²	11.63	1	11.63	2.34	0.1502	
Residual	64.63	13	4.97			
Lack of Fit	45.88	10	4.59	0.73	0.6902	not significant
Pure Error	18.75	3	6.25			
Cor Total	5788.68	27				

Table 6 Quadratic Model ANOVA for Final Setting Time

Source	Sum of Squares	Df	Mean Square	F Value	p-value Prob > F	
Model	18350.85	14	1310.78	100.34	< 0.0001	Significant
A-CEMII	840.50	1	840.50	64.34	< 0.0001	
B-CIC	4900.50	1	4900.50	375.13	< 0.0001	
C-CCW	3472.22	1	3472.22	265.80	< 0.0001	
D-W/B	392.00	1	392.00	30.01	0.0001	
AB	676.00	1	676.00	51.75	< 0.0001	
AC	1560.25	1	1560.25	119.44	< 0.0001	
AD	1225.00	1	1225.00	93.77	< 0.0001	
BC	4.00	1	4.00	0.31	0.5894	
BD	42.25	1	42.25	3.23	0.0954	
CD	2704.00	1	2704.00	206.99	< 0.0001	
A^2	131.28	1	131.28	10.05	0.0074	
B^2	671.45	1	671.45	51.40	< 0.0001	
C^2	1659.69	1	1659.69	127.05	< 0.0001	
D^2	180.53	1	180.53	13.82	0.0026	
Residual	169.82	13	13.06			
Lack of Fit	163.07	10	16.31	7.25	0.0650	not significant
Pure Error	6.75	3	2.25			
Cor Total	18520.68	27				

The F-value as shown in Table 5 & 6 is 82.23 and 100.34 for IST and FST respectively, with a value of P < 0.001 suggesting that the model is significant. The interactive effect of CIC and CCW is the most important relationship affecting the IST. The interactive effect of CCW and water-to-binder (w/b) ratio is a very relevant interaction that is affecting the FST (Table 5). Three-dimensional plot showing effect of CEMII and CIC on IST at a constant CCW and w/b ratio is presented in Figure 8a. The interaction of CEMII and CIC on IST indicated that IST increased gradually with a low CIC level because the CEMII uses the available water (limited) to form hydration products (Figure 8a). However, the trend became irregular as a reduction in IST was observed at higher CIC levels (17%-23%CIC). Similar trend was reported by [1]. The optimum IST could be obtained at about 56g of CIC, and at 256g of CEMII. There is a negative significant interaction between the CIC and CEMII.

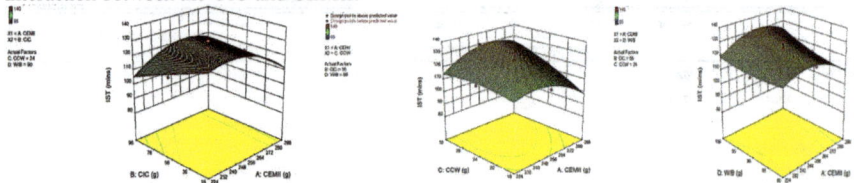

Figure 8 Interaction of (a) CIC & CEMII (b) CCW & CEMII (c) w/b & CEMII with IST

At a constant CIC and w/b ratio, IST continues to increase with higher levels of CCW (Figure 8b). The optimum IST could be obtained at about 24g of CCW, and at 256g of CEMII. [4] reported that the presence of CCW retards the hydration process of tricalcuim aluminates (Ca₃A) which is responsible for quick set. Combined effects of CIC and CCW pastes are important parameters as observed from their F-value of 324.24, being the highest values shown in Table 5. A suppressive effect was also observed by the two process variables. A constant amount of CIC and CCW (56g & 24g), minimum IST value of 115 minutes was observed against 288g of CEMII and w/b ratio of 0.28 (Figure 8c). However, the optimum IST factors observed were 256g of CEMII and 0.28. This trend corresponds with the results obtained by [2, 26]. There appears to be a limit to which water can be effective for binder development. The optimum FST could be obtained at about 56g of CIC, and at 256g of CEMII (Figure 9a). The interaction of CEMII and CCW on FST (Figure 9b) indicated that FST increased rapidly as CCW increased while CEMII decreased. This trend corresponds with the results obtained by [2]. Figure 9b show a synergistic effect between the process factors. The optimum factors are 256g of CEMII and 90g of water (w/b =0.28). Figure 9c indicated that the quantity of water above the optimum retards the final setting times. [2, 27] observed similar trends in their investigations.

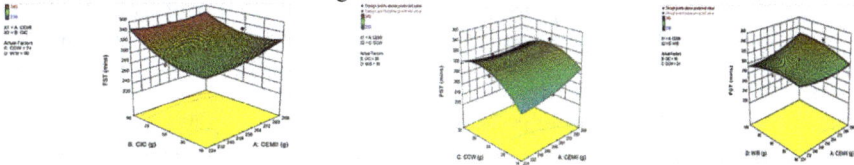

Figure 9 Interaction of (a) CIC & CEMII (b) CCW & CEMII (c) w/b & CEMII with FST

Optimization and validation

The optimum conditions obtained from this study are as follows: CEMII of 271.89g(77.1%), CIC of 61.062g(17.3%), CCW of 19.82g(5.6%), and water of 86.22g(24.4%). The corresponding optimized IST and FST are 114.705mins and 300.68mins respectively. At this combination, the Ca^{2+} needed to sustain the pozzolanic reaction is available, thus sustaining the formation of cementitious products toward achieving moderate setting times. Validation experimental runs were conducted and the average value of IST and FST were 118.65mins and 312.43mins respectively. In comparison with the predicted value, there is an error of about 3.44% and 3.91% for IST and FST; comparable with [22, 27].

Validation of the developed binder (CEMII-CIC-CCW) through Fourier Transform Infrared Spectroscopy

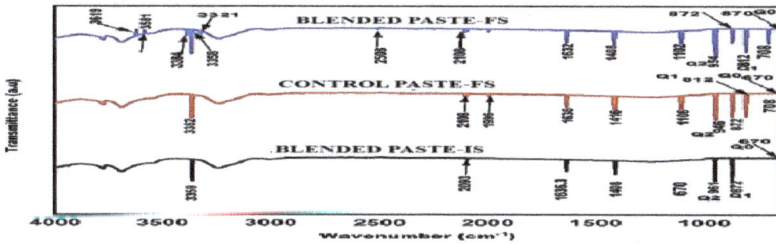

Figure 10 FTIR of control and blended pastes during setting periods

The blended pastes spectra showed some evident differences to the control pastes after setting (Figure 10; IS & FS) as a result of the additional aluminosilicate contributed by the CIC for pozzolanic reactions. First, the area ratio between the O-H portlandite $(Ca(OH)_2$ band confirmed that the $(Ca(OH)_2$ is higher in the blended paste with possible reaction at early stage due to activation by CCW [25, 28]. The characteristic vas H-O-H bands around $3200–3500cm^{-1}$ could be caused by adsorbed H_2O due to a higher surface area, hydration H_2O molecules trapped between the Ca-O layers of the C-S-H structure and the presence of Si-OH groups. The intensity of this band decreased as setting period increased. The most noteworthy change observed in the spectra was the presence of a band group associated to silica, centered at $946–954$ cm^{-1}; Si-O stretching vibrations (vas) of the tetrahedral units forming part of linear chains (Q2), present in the C-S-H structure [24, 25, 29].

Advances in Cement and Concrete
Materials Research Proceedings 51 (2025) 118-127

Materials Research Forum LLC
https://doi.org/10.21741/9781644903537-14

Conclusions

The individual and combined effects of some processing factors (water-binder ratio, % replacement of CIC and CEMII) on initial and final setting times of the developed binder have been examined by Central Composite Design model. The optimal factors were the weight percentages of 77.1% CEMII, 17.3% CIC, and 5.6% CCW, with a 0.244-w/b ratio and a desirability of 1. Under these conditions, the initial setting time was 118.65 minutes, the final setting time was 312.43 minutes. Based on the ANOVA test results, a reliable model has been developed to forecast the initial and final setting times of binders in building projects. The optimized material has the necessary pozzolanic binder properties, as indicated by characterization tests including FTIR.

References

[1] D. Snelson, S. Wild, M. O'Farrell, Setting times of Portland cement–metakaolin–fly ash blends, Journal of Civil Engineering and Management, 17, 1 (2011) 55-62.https://doi.org/ 10.3846/13923730.2011.554171

[2] P. R. J. Quiatchon, Investigation on the Compressive Strength and Time of Setting of Low-Calcium Fly Ash Geopolymer Paste Using Response Surface Methodology, Polymers, 13, 20, (2021) 3461. https://doi.org/ 10.3390/polym13203461

[3] N. Dave, A. K. Misra, A. Srivastava, S. Kaushik, Setting time and standard consistency of quaternary binders: The influence of cementitious material addition and mixing, International Journal of Sustainable Built Environment, 6, 1, (2017) 30-36. https://doi.org/ 10.1016/j.ijsbe.2016.10.004

[4] M. Babako, J. Apeh, Setting time and standard consistency of Portland cement binders blended with rice husk ash, calcium carbide and metakaolin admixtures, IOP Conference Series: Materials Science and Engineering, 805, 1, (2020) 012031. https://doi.org/ 10.1088/1757-899x/805/1/012031

[5] Y. Knop, A. Peled, Setting behavior of blended cement with limestone: influence of particle size and content, Materials and Structures, 49, 1 (2015) 439-452. https://doi.org/ 10.1617/s11527-014-0509-y

[6] M. H. Rashid, Strength Behavior of Cement Mortar Assimilating Rice Husk Ash.International Journal of Advances in Agricultural and Environmental Engineering, 3, 2,(2016) 45-48. https://doi.org/ 10.15242/ijaaee.a0416059

[7] E, E Ndububa, M. S. Omeiza, Potential of calcium carbide waste as partial replacement of cement in concrete, Nigerian Journal of Tropical Engineering, 9, 2, (2016) 1-9. ISSN: 1595-5397.

[8] H. Sun, Properties of Chemically Combusted Calcium Carbide Residue and Its Influence on Cement Properties, Materials, 8, 2, (2015) 638-651.https://doi.org/ 10.3390/ma8020638

[9] M. Al Salaheen et al., Modelling and optimization for mortar compressive strength incorporating heat-treated fly oil shale ash as an effective supplementary cementitious material using response surface methodology, Materials, 15, 19 (2022) 6538. https://doi.org/10.3390/ma15196538

[10] R. Kumar, Effects of high volume dolomite sludge on the properties of eco-efficient lightweight concrete: Microstructure, statistical modeling, multi-attribute optimization through Derringer's desirability function, and life cycle assessment, Journal of Cleaner Production, 307 (2021) 127107.https://doi.org/10.1016/j.jclepro.2021.127107

[11] K. Mermerdaş, Z. Algın, S. M. Olelwi, D. E. Nassani, Optimization of lightweight GGBFS and FA geopolymer mortars by response surface method,Construction and Building Materials, 139 (2017) 159–171. https://doi.org/10.1016/j.conbuildmat.2017.02.050

Advances in Cement and Concrete
Materials Research Proceedings 51 (2025) 118-127

Materials Research Forum LLC
https://doi.org/10.21741/9781644903537-14

[12] B. Şimşek, T. Uygunoğlu, H. Korucu, M. M. Kocakerim, Analysis of the effects of dioctyl terephthalate obtained from polyethylene terephthalate wastes on concrete mortar: A response surface methodology based desirability function approach application, Journal of Cleaner Production, 170 (2018), 437–445. https://doi.org/10.1016/j.jclepro.2017.09.176

[13] A. A. Raheem, M. A. Kareem, Chemical Composition and Physical Characteristics of Rice Husk Ash Blended Cement, International Journal of Engineering Research in Africa, 32 (2017), 25-35.https://doi.org/ 10.4028/www.scientific.net/jera.32.25

[14] R. Jaskulski, D. Jóźwiak-Niedźwiedzka, Y. Yakymechko, Calcined Clay as supplementary cementitious material, Materials, 13, 21 (2020) 4734.https://doi.org/10.3390/ma13214734

[16] D. Zhou, "Developing supplementary cementitious materials from waste London clay," Doctoral dissertation, Imperial College London, 2016.

[17] E. Atiemo, Studies on the effect of selected local admixtures on essential properties of cement for housing construction, (2012) Doctoral dissertation.

[18] A. S. Ogunro, M. A. Usman, E. E. Ikponmwosa, D. S. Aribike, Characterization of some clay deposits in Southwest Nigeria for use as Supplementary Cementitious Material in cement, in First International Conference on Advances in Cement and Concrete Research.

[19] O. Oribayo, A. P. Olalekan, R. U. Owolabi, O. O. Olaleye, O. A. Onyekaba, Adsorption of Cr(VI) ions from aqueous solution using rice husk–based activated carbon: optimization, kinetic, and thermodynamic studies, Environ Qual Manage, 4 (2020)1–17. https://doi.org/10.1002/tqem.2170

[20] R. U. Owolabi, M. A. Usman, A. J. Kehinde, Modelling and optimization of process variables for the solution polymerization of styrene using response surface methodology,Journal of King Saud University Engineering Sciences. http://dx.doi.org/10.1016/j.jksues.2015.12.005

[21] K. Scrivener, F. Martirena, S. Bishnoi, S. Maity, Calcined clay limestone cements (LC3), Cement and Concrete Research, 114 (2018) 49-56.

[22] BS EN 196-1 (2016). Methods of Testing Cement. Determination of Strength. British. Standard Institute. London

[23] BS EN 196 1995 Part 3: Methods of testing cement; determination of setting time and Soundness. British standard institution. London, 1995.

[24] J. Madejová, P. Komadel, Baseline Studies of the Clay Minerals Society Source Clays: Infrared Methods, Clays and Clay Minerals, 49, 5, (2001) 410–432.https://doi.org/ 10.1346/CCMN.2001.0490508

[25] H. Ez-zaki, J. M. Marangu, M. Bellotto, M. C. Dalconi, G. Artioli, L. Valentini, A Fresh View on Limestone Calcined Clay Cement (LC3) Pastes, Materials, 14, 11, (2021) 3037. https://doi.org/ 10.3390/MA14113037

[26] R. Zarzuela et al., Producing C-S-H gel by reaction between silica oligomers and portlandite: A promising approach to repair cementitious materials, Cement and Concrete Research, 130 (2020) 106008. https://doi.org/ 10.1016/J.CEMCONRES.2020.106008

[27] S. K. Antiohos, V. G. Papadakis, S. Tsimas, Rice husk ash (RHA) effectiveness in cement and concrete as a function of reactive silica and fineness, Cement and Concrete Research, 61 (2014) 20–27. https://doi.org/ 10.1016/J.CEMCONRES.2014.04.001

[28] Q. Wang, Y. Wang, X. Gu, J. Liu, X. Xu, Study on the Properties and Hydration Mechanism of Calcium Carbide Residue-Based Low-Carbon Cementitious Materials, Buildings, (2024). https://doi.org/ 10.3390/buildings14051259

Advances in Cement and Concrete
Materials Research Proceedings 51 (2025) 128-137

Materials Research Forum LLC
https://doi.org/10.21741/9781644903537-15

Development of an innovative and environmentally friendly alternative binder for application in the Nigerian construction industry: A review

Jacob Enejo ADEJO[1,a*], Usman MUAZU[2,b], Usman JAMILU[3,c] and Sarah GANDU[4,d]

[1]Department of Building, Lagos State University, Ojo, Lagos, Nigeria

[2]Department of Building, Federal Polytechnic, Kaura Namode, Kebbi, Nigeria

[3]Department Building, Ahmadu Bello University, Zaria, Kaduna, Nigeria

[4]Department of Glass and Silicate Technology, Ahmadu Bello University, Zaria, Kaduna, Nigeria

[a]jacobadejo21@gmail.com, [b]usmanradda@gmail.com, [c]jamilonline05@gmail.com, [d]srhgandu@gmail.com

Keywords: Alkali-Activated Binder, Alternative Binder, Construction Industry, Environmentally Friendly, Innovative, Portland Limestone Cement

Abstract. The cement-based construction industries particularly in developing countries like Nigeria faces numerous challenges amongst are natural resource depletion, energy consumption, production costs, and greenhouse gas emissions (GHG) that contribute to climate change. Therefore, the creation of an innovative and environmentally friendly binder that will serve as an alternative building material with sound adaptation and application in construction is imperative. Portland Limestone Cement (PLC) binders have been replaced by an alkali-activated binder (AAB) which appears to have better properties and environmental effects. This paper evaluated the state-of-the-art of alkali-activated binders, classification and reaction mechanism, constituent and factors influencing properties of AAB. In addition, it assessed the opportunities, application and sustainability of AAB. The discovery showed that the innovative and environmentally friendly binder promotes the advocacy of greenhouse construction. This is in line with sustainability requirements which is meeting our own needs without compromising the ability of future generations to meet their own needs. It is therefore recommended that alkaline-activated binders (AAB) should be used as an innovative and environmentally alternative binder to PLC for various building and civil engineering applications in the Nigerian construction industry.

Introduction

Over time, the global demand for cement manufacturing has expanded as a result of rapid infrastructural development. The increased production of cement has significantly increased the amount of greenhouse gases generated during its production [1]. Cement production requires a large amount of energy making the industry the third largest user of energy after the aluminum and steel industries [2, 3]. It is estimated every year that about 4 billion tonnes of Portland cement are produced worldwide with China dominating most of the market as shown in Figure 1. It has been estimated that for every 1 tonne of Portland limestone cement (PLC) produced, 0.5 –0.6 tonnes of CO_2 are released into the atmosphere. This atmospheric CO_2 generated contributes an annual 8% of all human CO_2 emissions. Additionally, the predicted contribution in 2050 has been projected to increase up to 26% of the total CO_2 anthropogenic emission if there are no significant changes in the existing cement production process as shown in Figure 2 [4,5,6] . Furthermore, the production of 1 ton of cement uses about 3.2GJ of natural energy and emits harmful greenhouse the environment, including carbon dioxide, nitrogen oxide and sulfur dioxide [7] .

Advances in Cement and Concrete
Materials Research Proceedings 51 (2025) 128-137

Materials Research Forum LLC
https://doi.org/10.21741/9781644903537-15

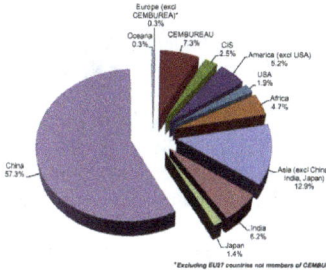

Figure 1: World Portland cement production today. Source: [4]

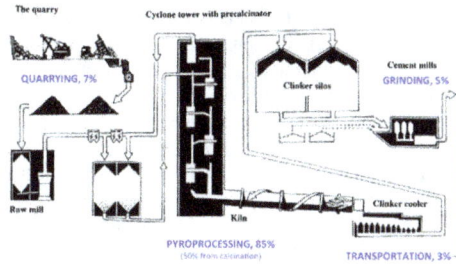

Figure 2: Share of total CO_2 emissions across the Portland cement production process. Source: [4]

In order to achieve sustainability goals, [8] explained that the construction industry, whose primary material is cement, must overcome a number of challenges. Some of these include the depletion of natural resources, energy consumption, production costs, and greenhouse gas emissions that cause climate change. To this effect, innovative, sustainable, and low carbon dioxide (CO_2) building materials emerged in an effort to reduce the environmental impact of the CO_2 produced as a result of the construction industry's use of Portland limestone cement (PLC) around the world. As these materials have the potential to support environmentally sustainable building processes [9] . These materials include kaolin, fly ash (FA), blast furnace slag (BFS), silica fume (SF), rice husk, groundnut shell, corn cob, sugarcane bagasse etc. They serve as supplementary cementitious materials (SCMs) and can be used to produce eco-efficient binders which are alternative to PLC. Example of such binders include calcium aluminate cements, calcium sulfo-aluminate cements, super-sulfated cements, geopolymers and alkali-activated binders [10].

History of alkaline activated binder
Historically, there have been many references to alkali activated materials before the 1940s but Glukhovsky was the first to extensively study the presence of analcime phases in the binders of ancient constructions. He later developed binders made from aluminosilicates in reaction with alkaline industrial wastes, which he named "soil silicate concrete" and "soil cements". The early investigations on alkali-activated binders mainly focused on the activation of blast furnace slag, a by-product of the metallurgical industry. Also, Davidovits developed and patented a novel binder which he named "geopolymer cement". The first geopolymer binder was a slag-based geopolymer cement which consisted of metakaolin, blast furnace slag and alkali silicate [11,12]. The benefit of this technology compared to PLC technology is that it hardened faster while reaching the maximal strength at early hours and provide a durable and compact microstructure.

Classification of alkali-activated binders
Alkali activated binder (AAB) is the broadest classification, encompassing essentially any binder system derived by the reaction of alumino-silicate material with or without calcium oxide (CaO) and an alkaline activator [13]. The term "alkali-activated materials", and "geopolymer" found in literature, are mostly used interchangeably to describe alkali activated binders. However, a clear difference between them is based on the high calcium content in the raw materials for alkali-activated materials and low-calcium content materials for geopolymers [14]. Notwithstanding, AAB with high or low content of calcium is produced by the chemical reactions of precursors (containing silica oxide (SiO_2), alumina oxide (Al_2O_3), and calcium oxide (CaO) contents, with

alkaline activators. The precursors could be industrial by-products like fly ash, the material of geological origin like metakaolin (MK), and agricultural waste like groundnut shell ash (GSA) [15,16]. The alkaline activator comprises; sodium or potassium silicate (Na_2SiO_3 or K_2SiO_3) and sodium or potassium hydroxide (NaOH or KOH).

Alkali-activated binders reduce CO_2 emissions primarily by eliminating calcium carbonate ($CaCO_3$) precursors and high temperatures for processing PLC [17]. In addition, AAB possess qualities that are comparable to those of PLC, but they have a substantially less carbon footprint and may perform better than conventional cements in some situations [9]. In fact [18], opined that AAB could reduce CO_2 emission associated with PLC production by 80%. Therefore, the use of locally available raw materials to produce AAB, with well-formulated mix designs, and an efficient production process result in construction materials that are both affordable and environmentally friendly.

Reaction Mechanism of Alkaline Activated Binder
It involves dissolving the aluminosilicate material in an alkaline liquid, transporting the dissolved units, condensing the units into a polymeric network, and then hardening the network. The system becomes more alkaline (pH increases) during the exothermic dissolution phase, which causes the aluminosilicate solids to break down into unstable, reactive units made up of covalently bonded Si-O-Si and Al-O-Si units. After that, additional phases are created by Al atoms penetrating Si-O-Si compound, which results in the development of aluminosilicate gels (colloid phase) and a coagulated structure. Then crystallization and the creation of a condensed structure occur [13].

Based on the composition of the product, AAB can be classified into two types of reaction mechanism which forms an inorganic binder through a polymerization process. The first system involves the activation of calcium-rich raw materials like blast furnace slags, with high content of Si, Al and Ca atoms. In this case, reaction products contain primarily silicate and calcium and hydroxide solutions. The High content of calcium-rich aluminosilicates thus lead to the formation of calcium aluminosilicate hydrate (C-A-S-H) gels. Similarly, the second type of the reaction mechanism of AAB, involves the activation of low calcium aluminosilicate material using high alkaline solution. This reaction results to the formation of amorphous zeolitic phase with high mechanical properties similar or better than PLC. The alkali-activation of low calcium aluminosilicates leads to formation of so-called sodium aluminosilicate hydrate (N-A-S-H) gels [11,19,14]. Figure 3 shows a short illustration of the reaction mechanism of alkali-activated binders in terms of characteristics of the raw materials, activating solutions and reaction mechanisms.

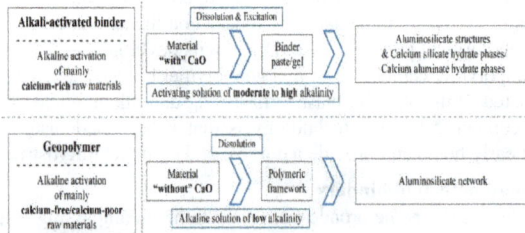

Figure 3: Reaction Mechanisms of Alkali-Activated Binder. Source: [14]

Constituents of Alkali-Activated Binders
The choice for selection of source material for making AAB depends on factors such as availability, cost, type of application, and specific demand of the end user. Basically, the two key ingredients of AAB are aluminosilicate source material and alkaline activator and aluminosilicate sources as presented in Figure 4.

Advances in Cement and Concrete
Materials Research Proceedings 51 (2025) 128-137

Materials Research Forum LLC
https://doi.org/10.21741/9781644903537-15

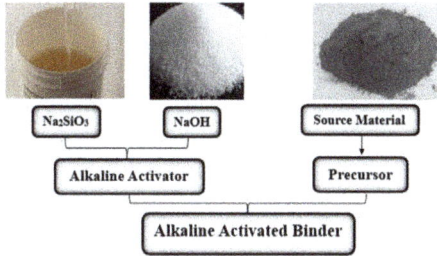

Figure 4: Constituents of Alkali-Activated Binders

Aluminosilicate source materials (precursor)

Fly ash

Fly ash is an industrial by-product obtained from the burning of bituminous coal and is considered suitable as source material for AAB. It has been estimated that the annual production of fly ash is one billion tons, which can introduce environmental problems if not stored and disposed of properly. Fly ash is an acidic material containing acidic oxides such as Al_2O_3, SiO_2, and Fe_3O_2 which provide the potential for alkaline reaction. Most of the fly ash utilized as a source for aluminosilicate belong to Class F pozzolana. However, due to its quicker setting and availability, high calcium fly ash, also known as Class C, is not frequently employed as an alkaline binder's precursor [1].

Metakaolin

Metakaolin can be described as a dehydroxylated pozzolanic product derived from the high-temperature firing of raw kaolin. Kolin or kaolinite ($Al_2Si2O_5(OH)_4$) is a clay mineral that contains a high amount of layered tetrahedral silicon atoms connected via oxygen to octahedral aluminum atoms [20]. Metakaolin is a highly reactive supplementary cementitious material rich in alumina and silica produced by calcining kaolinite clay in the range from 600°C to 900°C [21]. During the calcination process, the chemically bound water in the kaolinite clay evaporates through the dehydroxylation process and breaks down the raw material's structure, and eventually forms an amorphous phase known as metakaolin.

Groundnut shell ash (GSA)

According to [22], groundnut shell ash (GSA) is an ash created by burning groundnut shell (GS) at a temperature between 500 - 650°C. [23] stated that it contains significant amounts of silica, alumina, and calcium oxide, making it a potential precursor for the synthesis of AAB. GS is an agricultural waste product of groundnut milling. Over time, the shell has come to be thought of as one of the more typical solid wastes, particularly in developing nations [24]. As of 2019, there were approximately 48.76 million metric tons (mt) of groundnuts produced worldwide and of this number, 25% to 40% of the groundnuts are shell, which is an approximation of the amount of solid waste that will be created [25]. The shell can be used to produce animal feed, but because it has a high calorific value, it can also be burned to produce energy. The ash produced after combustion is often dumped in an open area or a landfill because it has no use in the economy. The natural environment is polluted by the ash disposal.

Advances in Cement and Concrete   Materials Research Forum LLC
Materials Research Proceedings 51 (2025) 128-137   https://doi.org/10.21741/9781644903537-15

Alkaline Activator

This alkaline activator contains alkali cations that quickens the precipitation and crystallization of the siliceous and aluminous specie present in the aluminosilicate source materials to form AAB. According to [25], a combination of sodium hydroxide (NaOH) or Potassium hydroxide (KOH) and Sodium Silicate (Na_2SiO_3) or Potassium Silicate (K_2SiO3) can be used as the alkaline activator. However, NaOH and Na_2SiO_3 are the most commonly used ones for the production of AAB. According to [20], polymerization reaction is stirred by the ability of alkaline solution to dissolve the aluminosilicate present in the aluminosilicate source materials and release the reactive silicon and aluminum into the solution to form polymeric silicon-oxygen-aluminum bonds (Si-O-Al).

Sodium hydroxide (NaOH)

Sodium hydroxide (NaOH)/Potassium hydroxide (KOH) are alkali activators that can be used for the production of geopolymer. They help the dissolution of the aluminosilicate source. The dissolution of aluminosilicate sources increases with increasing concentration of alkali activators. This dissolution ability of the aluminosilicate source is responsible for the ultimate strength of AAB. NaOH is commonly used for the production of AAB because it promotes better polymerization than KOH due to the fact that aluminosilicate source materials are more easily dissolved in NaOH solution than in KOH solution [26, 27].

Sodium silicate (Na_2SiO_3)

Sodium silicate is the generic name for a series of compounds derived from soluble sodium silicate glasses. It is a colourless compound of oxides of sodium and silicon highly miscible in water. It has a series of chemical formulas varying in sodium oxide (Na_2O) and silicon dioxide or silica (SiO_2) contents. The most important property of sodium silicate is the weight ratio of SiO_2: Na_2O. It is commercially produced in the ratio range of 1.5 to 3.2. The varying proportions of SiO_2 to Na_2O and the solids content results in solutions with differing properties that have many diversified industrial applications. For AAB production, sodium silicate is mixed with NaOH to enhance alkalinity and increase the strength of AAB. According to [28], the creation of geopolymer with high strength and durable geopolymer requires sodium silicates with a molar ratio of SiO_2:$N_2O>1.65$.

Molarity Calculation for Sodium Hydroxide (NaOH)

According to [29] the role of NaOH and KOH in AAB is that it determines the number of ions and controls the dissolution process of the aluminosilicate source material. It should be noted that the concentration of NaOH solution can vary from 6 to 16 molar. However, [30] explained that the results of past studies on the molarity of NaOH have shown that an increase in its molarity results in increased strength of AAB. However, an excess concentration of NaOH has the tendency to reduce the effectiveness of AAB's fresh and hardened properties. Most importantly, the addition of NaOH to the alkaline activator considerably improved the characteristics of AAB. Before using sodium hydroxide (NaOH) solids it should be dissolved in water with the required concentration, the concentration of NaOH solution can vary from 8 to 16 molar. The mass of NaOH solids in a solution varies depending on the concentration of the solution; for example, NaOH solution with a concentration of 8 molar consists of $8 \times 40 = 320g$, $10 \times 40 = 400g$, $12 \times 40 = 480g$, $14 \times 40 = 560g$ and $16 \times 40 = 640g$ of NaOH solids per liter of the solution, where 40 is the molecular weight of NaOH. The mass of NaOH solids was measured as 262g per kg of NaOH solution with a concentration of 8 molar, the mass of NaOH solids per kg of the solution for other concentrations has been measured as 10 molars for 314g, 12 molars for 361g, 14 molars for 404g, and 16 molars

Advances in Cement and Concrete   Materials Research Forum LLC
Materials Research Proceedings 51 (2025) 128-137   https://doi.org/10.21741/9781644903537-15

for 444 g [31]. It should also be noted that the mass of NaOH solids is only a fraction of the mass of the NaOH solution, and water is the major component.

Effect of Curing Temperatures

Curing of AAB is said to be beneficial as it increases the reaction rate by accelerating the dissolution of silicon (Si), aluminum (Al) and calcium (Ca) contained in the precursor as well facilitate the polycondensation process and hardening of AAB matrix. According to [30], the curing conditions have a major impact on the polymerization mechanism, the quality of the polymeric product, and the development of strength. AAB made with various precursor materials are cured using ambient temperature, elevated, steam, and water ponding. Heat curing involves subjecting AAB specimens to a higher temperature for faster curing. The specimens are subjected to temperatures around 60°C for 24 h in an oven after demoulding them. While for steam curing it is done by subjecting the specimens to the water vapour under atmospheric or high-pressure conditions. Steam curing is commonly performed in the temperature range of 70°C - 100°C. Even though this curing procedure uses a lot of energy, it is frequently employed for AAB specimens where the polymerization reaction needs a lot of external energy to start. Furthermore, the precast concrete industry frequently uses this curing procedure to achieve strength quickly and remove form-work early. The curing temperature was raised from 20 °C to 60 °C, the compressive strengths of AAB and setting time decreased as shown in figures 4 and 5. Nevertheless, compressive strength decreases at 80 °C and 100 °C, because of fast/flash setting prevents the AAB product from becoming a compact and durable structure [32, 33, 34, 35, 36, 37].

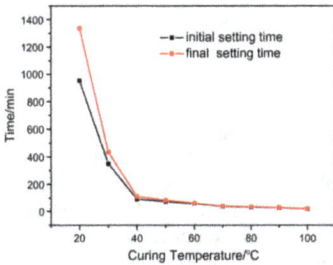

Figure 5: Setting Time of Alkali Activated Binder Cured at Varying Temperature. Source: [36]

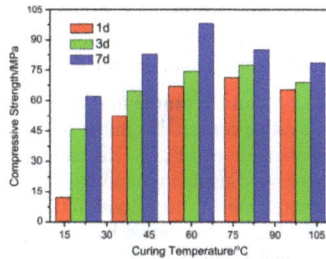

Figure 6: Compressive Strength of Alkali Activated Binder Cured at Varying Temperature. Source: [36]

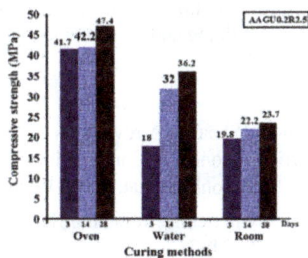

Figure 7: Compressive Strength of Alkali Activated Binder Cured at Varying Temperature. Source: [37]

Sustainability of alkaline activated binder

From the environmental viewpoint, alkaline activated successfully develops and improves the concept of greenhouse construction in accordance with sustainability standard. Without Portland cement, alkali-activated binders (AAB) can lower CO_2 emissions [38]. AAB is produced using waste, including metakaolin from geological origin, residual ashes from thermal power plants, such as Fly Ash, and agricultural waste, such as groundnut shell ash (GSA). Materials with a low carbon footprint make up the trash kind. When compared to PLC, the mechanical and durability attributes of AAB made from these wastes are outstanding, on par with or even better [17].

According to [39] it is very difficult to estimate the amount of agricultural waste produced annually. This is because as the population continues to increase, their number continues to increase. The low price of these wastes and their economic values are less than the cost of gathering, transportation, and processing them for advantageous use. This means that most times, these wastes are incinerated. However, upon incinerating agricultural waste, the resulting ash are disposed in landfills and in most cases, it degrades the environment and also creates problems which leads to the threat of intoxication to the humans and animals [40]. The ash residue obtained from the incineration of these agricultural waste when used as fuel contains oxide like SiO_2, Al_2O_3 and CaO that qualifies them to be used as supplementary cementitious materials (SCM) [41].

Application of alkali activated binder

Despite the numerous challenges that AAB encounters, particularly in production, there are still several applications for it. These applications include reinforced concrete, plain concrete, precast concrete components, both reinforced and unreinforced concrete (such as pipes), as well as mortars, grouts, and renders. Additionally, AAB can be used to create foamed and lightweight concrete, recycled aggregate concrete (RCA) as well as matrices for the immobilization of toxic and nuclear wastes, both organic and inorganic [42,43]

Conclusion and Recommendation

The innovative perspective in the building industry is that alkali-activated binders will present a greener alternative to PLC. This paper focused on the development of alkali-activated binders as an innovative construction and building material. Historical background, classification, reaction mechanisms, constituent, opportunities, and application of alkaline activated binder with how it is sustainable as an innovative building and construction material was examined. Also, the exact reaction mechanism of alkali-activated binders is not yet quite understood, because it depends on the precursor and on the alkaline activator. However, for the Nigerian construction industry, adopting alkali-activated binders offers a revolutionary possibility. By lowering energy use, greenhouse gas emissions, and reliance on conventional, resource-intensive materials, these alkali-activated binders support sustainable development goals. Their substantial environmental advantages such as enhanced mechanical and durability properties will work together to increase the longevity and efficiency of building structures. It is therefore recommended that alkaline-activated binders should be used as an alternative to PLC for various building and civil engineering applications in the Nigerian construction industry.

References

[1] M. M. A. Elahi, M. M. Hossain, M. R. Karim, M. F. M. Zain, C. Shearer, A review on alkali-activated binders: Materials composition and fresh properties of concrete, Construction and Building Materials, (2020), 260, 119788. https://doi.org/10.1016/j.conbuildmat.2020.119788

[2] S. J. Herbert, N. Sakthieswaran, B. G. Shiny, Review on geopolymer concerte with different additives. International Journal of Engineering Research, (2015), 1(2), 21–31.

Advances in Cement and Concrete
Materials Research Proceedings 51 (2025) 128-137

Materials Research Forum LLC
https://doi.org/10.21741/9781644903537-15

[3] A. M, Robbie, Global CO2 emissions from cement production , 1928 – 2017 1 Introduction to Previous estimates of global cement emissions. Earth System Science Data, (2021), 10(4), 1–20.

[4] M. S. Imbabi, C. Carrigan, S. Mckenna, Trends and developments in green cement and concrete technology. International Journal of Sustainable Built Environment, (2013), 1(2), 194–216. https://doi.org/10.1016/j.ijsbe.2013.05.001

[5] K. Arbi, M. Nedeljković, Y. Zuo, G. Ye, A Review on the Durability of Alkali-Activated Fly Ash/Slag Systems: Advances, Issues, and Perspectives. Industrial and Engineering Chemistry Research, (2016), 55(19), 5439–5453. https://doi.org/10.1021/acs.iecr.6b00559

[6] I. Pol Segura, N. Ranjbar, A. Juul Damø, L. Skaarup Jensen, M. Canut, P. Arendt Jensen, A review: Alkali-activated cement and concrete production technologies available in the industry. Heliyon, (2023), 9(5). https://doi.org/10.1016/j.heliyon.2023.e15718

[7] P. Devarajan, D. S. Vijayan, S. R. Kumar, A. F. J. King, A short review on Substantial role of Geopolymer in the sustainable construction industry. IOP Conference Series: Earth and Environmental Science, (2023), 1130(1). https://doi.org/10.1088/1755-1315/1130/1/012002

[8] H. S. Gökçe, M. Tuyan, M. L. Nehdi, Alkali-activated and geopolymer materials developed using innovative manufacturing techniques: A critical review. Construction and Building Materials, (2021), 303. https://doi.org/10.1016/j.conbuildmat.2021.124483

[9] A. Attanasio, L. Pascali, V. Tarantino, W. Arena, A. Largo, Alkali-activated mortars for sustainable building solutions: Effect of binder composition on technical performance. Environments - MDPI, (2018), 5(3), 1–14. https://doi.org/10.3390/environments5030035

[10] M. C. G. Juenger, F. Winnefeld, J. L. Provis, J. H. Ideker, Advances in alternative cementitious binders. Cement and Concrete Research, (2011), 41(12), 1232–1243. https://doi.org/10.1016/j.cemconres.2010.11.012

[11] C. Shi, K. V. Pavel, D. Roy, Alkaline-Activated Cements and Concrete. Taylor and Francis, CRC Press, (2006). https://doi.org/10.1201/9781482266900

[12] M. T. Marvila, A. Rangel, G. D. Azevedo, Reaction mechanisms of alkali-activated materials. IBRACON Structures and Materilas Journal, (2021), 14(3). https://doi.org/10.1590/S1983-41952021000300009

[13] F. Pacheco-Torgal, J. Castro-Gomes, S. Jalali, Alkali-activated binders: A review. Part 1. Historical background, terminology, reaction mechanisms and hydration products. In Construction and Building Materials (2008), 22(7),1305–1314. https://doi.org/10.1016/j.conbuildmat.2007.10.015

[14] V. B. Thapa, D. Waldmann, A short review on alkali-activated binders and geopolymer binders. (n.d.) 2–4.

[15] A. A. Adam, Strength and Durability Properties of Alkali Activated Slag and Fly Ash-Based Geopolymer Concrete (Issue August), RMIT University Melbourne, Australia, (2009).

[16] M. Torres-Carrasco, F. Puertas, Activarea Alcalină a Unor Aluminosilicaţi Ca Alternativă La Cimentul Portland : Review Alkaline Activation of Aluminosilicates As an Alternative To Portland Cement: a Review. Revista Română de Materiale / Romanian Journal of Materials, (2017), 47(1), 3–15.

[17] J. Payá, L. Soriano, A. Font, M. Victoria, B. Rosado, J. A. Nande, J. Mar, M. Balbuena, Reuse of Industrial and Agricultural Waste in the Fabrication of Geopolymeric Binders : Mechanical. Materials, (2021), 14.

[18] J. S. J. Van Deventer, J. L. Provis, P. Duxson, Technical and commercial progress in the adoption of geopolymer cement. Minerals Engineering, (2012), 29, 89–104. https://doi.org/10.1016/j.mineng.2011.09.009

[19] M. Hongqiang, C. Hongyu, Z. Hongguang, S. Yangyang, N. Yadong, H. Qingjie, H., Zetao, Study on the drying shrinkage of alkali-activated coal gangue-slag mortar and its mechanisms. Construction and Building Materials, (2019), 225, 204–213. https://doi.org/10.1016/j.conbuildmat.2019.07.258

[20] J. C. Petermann, A. Saeed, H. I. Michael, Alkali Activated Material: a Literature Review. (2012).

[21] A. T. Bakera, M. G. Alexander, Use of metakaolin as a supplementary cementitious material in concrete, with a focus on durability properties. RILEM Technical Letters, (2019), 4, 89–102. https://doi.org/10.21809/rilemtechlett.2019.94

[22] J. G. D. Nemaleu, V. Bakaine Djaoyang, A. Bilkissou, C. R. Kaze, R. B. E. Boum, J. N. Y. Djobo, P. Lemougna Ninla, E. Kamseu, Investigation of Groundnut Shell Powder on Development of Lightweight Metakaolin Based Geopolymer Composite: Mechanical and Microstructural Properties. Silicon, (2020). https://doi.org/10.1007/s12633-020-00829-z

[23] N. V. Lakshmi, P. S. Sagar, Experimental study on determination of compressive strength of ground nut shell ash on partial replacement with cement. International Journal for Research in Applied Science and Engineering Technolog, (2017), 5(Viii), 1492–1496. https://doi.org/10.22214/IJRASET.2017.8211

[24] R. Priya, P. Partheeban, Durability Study of Low Calcium Flyash Based Geopolymer Concrete. Indian Journal of Applied Research, (2013).

[25] J. Usman, N. Yahaya, E. M. Mohammed. Influence of groundnut shell ash on the properties of cement, IOP Conf. Series: Materials Science and Engineering, 2019, 601 (012015). https://doi.org/10.1088/1757-899X/601/1/012015

[26] Liew, Y. M., Heah, C. Y., Mohd Mustafa, A. B., & Kamarudin, H. (2016). Structure and properties of clay-based geopolymer cements: A review. Progress in Materials Science, 83, 595–629. https://doi.org/10.1016/j.pmatsci.2016.08.002

[27] D. D. Burduhos Nergis, M. M. A. B. Abdullah, P. Vizureanu, M. F. Mohd Tahir, Geopolymers and Their Uses: Review, IOP Conference Series: Materials Science and Engineering, (2018), 374(1). https://doi.org/10.1088/1757-899X/374/1/012019

[28] J. Davidovits, Geopolymer Cement a review, Geopolymer Science and Technics, (2013). 0, 1–11. www.geopolymer.org

[29] X. Guo, H. Shi, L. Chen, W. A. Dick, Performance and Mechanism of Alkali-Activated Complex Binders of High-Ca Fly Ash and Other Ca-Bearing Materials, Journal of Hazardous Materials. (2010), 173(1-3):480-6. https://doi.org/10.1016/j.jhazmat

[30] M. Ibrahim, M. Maslehuddin, An overview of factors influencing the properties of alkali-activated binders. Journal of Cleaner Production, (2021) 286, 124972. https://doi.org/10.1016/j.jclepro.2020.124972

Advances in Cement and Concrete
Materials Research Proceedings 51 (2025) 128-137

Materials Research Forum LLC
https://doi.org/10.21741/9781644903537-15

[31] D. Hardjito, B. V. Rangan, Development and properties of Low- Calcium Fly Ash Based Geopolymer Concrete. Research Report GC 1, 1-94, Faculty of Engineering, Curtin University of Technology, Perth, Australia. (2005), Available at espace@curtin or www.geopolymer.org. Accessed 23/11/2024.

[32] S. Aydin, B. Baradan, Mechanical and microstructural properties of heat cured alkali-activated slag mortars. Materials and Design, (2012), 35, 374–383. https://doi.org/10.1016/j.matdes.2011.10.005

[33] L. P. Qian, Y. S. Wang, Y. Alrefaei, J. G. Dai, Experimental study on full-volume fly ash geopolymer mortars: Sintered fly ash versus sand as fine aggregates. Journal of Cleaner Production, (2020), 263, 121445. https://doi.org/10.1016/j.jclepro.2020.121445

[34] A. Bilginer, O. Canbek, S. Turhan Erdoğan, Activation of Blast Furnace Slag with Soda Production Waste. Journal of Materials in Civil Engineering, (2020), 32(1), 1–9. https://doi.org/10.1061/(asce)mt.1943-5533.0002987

[35] G. Yıldırım, A. Kul, E. Özçelikci, M. Şahmaran, A. Aldemir, D. Figueira, A. Ashour, Development of alkali-activated binders from recycled mixed masonry-originated waste. Journal of Building Engineering, (2021), 33. https://doi.org/10.1016/j.jobe.2020.101690

[36] B. H. Mo, H. Zhu, X. M. Cui, Y. He, S. Y. Gong, Effect of curing temperature on geopolymerization of metakaolin-based geopolymers. Applied Clay Science, (2014), 99, 144–148. https://doi.org/10.1016/j.clay.2014.06.024

[37] T. O. Yusuf, M. Ismail, J. Usman, A. H. Noruzman, Impact of Blending on Strength Distribution of Ambient Cured Metakaolin and Palm Oil Fuel Ash Based Geopolymer Mortar. Advances in Civil Engineering, (2014). https://doi.org/10.1155/2014/658067

[38] M. Pavlin, K. König, J. König, U. Javornik, V. Ducman, Sustainable Alkali-Activated Slag Binders Based on Alternative Activators Sourced From Mineral Wool and Glass Waste. Frontiers in Materials, (2022), 9. https://doi.org/10.3389/fmats.2022.902139

[39] T. T. Nguyen, C. I. Goodier, S. A. Austin, Factors affecting the slump and strength development of geopolymer concrete. Construction and Building Materials, (2020), 261. https://doi.org/10.1016/j.conbuildmat.2020.119945

[40] V. K. Singh, & G. Srivastava, . Development of a sustainable geopolymer using blast furnace slag and lithium hydroxide. Sustainable Materials and Technologies, (2024), 40. https://doi.org/10.1016/j.susmat.2024.e00934

[41] D. A. Oliveira, P. Benelli, E. R. Amante, A literature review on adding value to solid residues: Egg shells. Journal of Cleaner Production, (2013), 46, 42–47. https://doi.org/10.1016/j.jclepro.2012.09.045

[42] J. E. Adejo, E. T. Ka'ase, D. D. Dahiru, M. M. Garba, Durability Properties of Metakaolin Based Geopolymer Concrete Made with Recycled Concrete Aggregate, Journal of Health and Environmental Studies, (2017), 1(1&2), 230-237. SSN 2323-163X. www.unimaid.edu.ng

[43] J. Adejo, E. Ka'ase, M. Muhammad, (2020). Properties of Metakaolin Based Geopolymer Concrete Made With Recycled Concrete Aggregate. International Journal of Research and Innovation in Applied Science (IJRIAS), (2020), V(IX), 57–63. www.rsisinternational.org

Advances in Cement and Concrete
Materials Research Proceedings 51 (2025) 138-145

Materials Research Forum LLC
https://doi.org/10.21741/9781644903537-16

Effect of water/cement ratio and delay in casting on the workability and compressive strength characteristics of laterized concrete

Joseph Adurapemi OLUSOLA[1,a*], Kolapo Olubunmi OLUSOLA-ELEKA[2,b],
Isaac Taiwo AFUYE[1,c], Adetomi Oluwabukunmi ABEGUNDE[2,d],
Kazeem Dele MUSBAU[2,e]

[1]Department of Building, Bamidele Olumilua University of Education, Science and Technology, Ikere-Ekiti, Ekiti State, Nigeria

[2]Department of Building, Osun State University, Osogbo, Osun-State, Nigeria

[a]olusola.joseph@bouesti.edu.ng, [b]kolapo.olusola-eleka@uniosun.edu.ng, [c]afuye.isaac@bouesti.edu.ng, [d]adetomi.abegunde@uniosun.edu.ng, [e]kazeem.musbau@uniosun.edu.ng

Keywords: Compressive Strength, Workability, Water/Cement Ratio, Delay in Casting, Laterized Concrete

Abstract. This paper reports the result of investigations on the effect of water cement ratio and delay in casting on the workability and compressive strength characteristics of laterized concrete. Three mix ratios 1:1:2, 1:1.5:3, 1:2:4 and varying fine aggregate fraction replacements of sand with laterite amounting to 0%, 25%, 50%, 75% and 100% were investigated. Test result showed that the workability increases with increasing water cement ratio and decreases with increasing period of delay of casting. The compressive strength decreases with increasing water cement ratio. Delaying the casting for up to 40 minutes for 1:1:2 mix, 30 minutes for 1:1.5:3 mix and between 15 to 25 minutes for 1:2:4 mix resulted in beneficial effects amounting averagely to 20% gain in compressive strength with no appreciable loss in workability. The use of superplasticizers to improve workability and mechanical compaction to improve strength in laterized concreting processes is recommended.

Introduction

Laterized concrete is concrete in which the sand content has partially or wholly been replaced by laterite fines, a clayey-sand cohesive soil [1]. The dire need towards development, optimal and sustainable utilisation, and improvement of indigenous material and construction technology has prompted research activities on the practical usefulness of laterite and its combinations with other types of aggregates in concrete production.[2]. Previous studies have shown that source of laterite, richness of mix proportions including laterite/cement ratio, method of curing, aggregate sizes, curing age, water/cement ratio, balancing workability with strength objectives and processing of laterite before use are significant factors in the characteristic strength of laterized concrete [3-10]. The present research investigated effect of water/cement ratio and delay of casting on workability and compressive strength characteristics of laterized concrete having varying laterite contents.

Materials and Methods

Materials

The basic materials used in this study were laterite, river sand, both of which were used as fine aggregates and have maximum particle size of 2.36 mm, locally sourced washed gravel used as coarse aggregate with sizes ranging from 5 to 20 mm, Portland Limestone Cement (PLC) whose properties conform with the requirements of [11] for Portland Cement used as binder and water obtained from a nearby flowing stream. The lateritic soil was collected from a borrow pit at a depth of 1.2m along Ife-Ifewara Road in Ife Central Local Government Area of Osun State of Nigeria,

Advances in Cement and Concrete
Materials Research Proceedings 51 (2025) 138-145

Materials Research Forum LLC
https://doi.org/10.21741/9781644903537-16

at a location which falls within Latitude 7°49'40.76" and Longitude 4°48'22.62''. River sand was collected from Osun River basin.

Methods
Preparation of test samples
Three commonly encountered nominal mix ratios namely 1:1:2, 1:1.5:3 and 1:2:4 (and listed in order of decreasing richness) involving laterite as substitutes for sand in the fine aggregate fraction in gradation of 0%, 25%, 50%, 75% and 100% were used in each case. The research results reported by [5] in which laterite was used wholly as fine relates the laterite/cement ratio (Y) to the optimum water/cement ratio (X) as shown in Equation 1

$$Y = -0.9 + 38.5X \dots\dots\dots (1)$$

Based on Equation 1, the optimum water cement ratios for mix proportions 1:1:2. 1:1.5:3 and 1:2:4 were calculated as being approximately equal to 0.50, 0.65 and 0.75. The present work involves two tests classified as Main Tests (MT) and Supplementary Tests (ST). In the main test, the water cement ratio was varied around the aforementioned calculated "optimum" values in equal steps. In the supplementary tests, values of water/cement ratios higher than the maximum adopted in the main test were investigated. The wide range of water/cement ratios investigated in both cases and shown in Table 1 was expected to give a clearer understanding of the effects of water content on the compressive strength and workability characteristics of laterized concrete.

Table 1 *Mix proportion and corresponding water/cement ratios (by weight)*

Mix 1:1:2		Mix 1:1.5:3		Mix 1:2:4	
MT	ST	MT	ST	MT	ST
0.40	0.65	0.55	0.80	0.65	0.90
0.45	0.70	0.60	0.85	0.70	0.95
0.50	0.75	0.65	0.90	0.75	1.00
0.55	0.80	0.70	0.95	0.80	1.05
0.60	0.85	0.75	1.00	0.85	1.10

Tests applied to the samples
The concrete mixes were prepared, cast into 150mm cube moulds, cured for 28 days and tested respectively in accordance with the provisions of [12] for workability and [13] for compressive strength. The properties of the materials as listed in Tables 2 and 3 were determined in compliance with established standards. In investigating the effect of delay in casting on workability and compressive strength of laterized concrete mixes, a fixed mixing duration of 5 minutes was adopted. The zero-minute delay in casting implies slump test was carried out immediately after mixing.

Results and Discussion
Physical properties
Table 2 summarizes the physical properties of the soil samples used. The calculated values of the fineness moduli of the lateritic soil and sand samples indicated fine aggregates of medium grading. All the soils are uniformly graded (coefficient of uniformity between 1.0 and 5.0) and hence are suitable for producing good concrete. The values obtained from Atterberg's limit tests indicated that laterite used is in the class of very clayey-sand cohesive soil of intermediate plasticity (LL between 35% and 50%, PI between 20% and 30%) [1, 7].

Advances in Cement and Concrete
Materials Research Proceedings 51 (2025) 138-145

Materials Research Forum LLC
https://doi.org/10.21741/9781644903537-16

Table 2 Summary of the physical properties of soil samples used

Properties	Types of Soil		
	Laterite	Sand	Gravel
Fineness modulus	3.03	3.12	6.80
Coefficient uniformity	4.96	4.42	2.50
Relative density	2.55	2.64	2.66
Initial (as received) moisture content (%)	14.71	6.32	1.86
Moisture content (as used) (%)	8.71	3.50	1.29
Liquid limit (LL.%)	36.50	-	-
Plastic Limit (PL.%)	17.50	-	-
Plasticity index (PI.%)	19.00	-	-

Chemical analysis of lateritic soil
Table 3 shows the results of the analysis of the chemical composition of a sample of the lateritic soil used. The values are comparable to those earlier obtained by Adepegba [3]. Silica (SiO_2) is observed to form about 63% of laterite while for the white river sand, this mineral has been observed, from literature [16], to form 99% of its chemical composition. This implies that river sand is almost a pure quartz material (a stiff, stable, lowly porous having a high compressive strength). The comparatively higher quartz content of river sand might partly have been responsible for higher compressive strength recorded in previous studies for normal concrete when compared with laterized concrete from same mix proportions. Table 3 further reveals that the Loss on Ignition (LOI) for the soft laterite sample is close to 9%. For white sand, the corresponding value is often less than 1% [16]. These values indicate that river sand is much cleaner than laterite. This further reinforced the significance of processing laterite before use vis-à-vis its resulting effect on the strength properties of laterized concrete [8]

Table 3 Chemical composition of lateritic soil sample used

Mineral Constituent	Percentage Composition (%)	
	Present Work	Previous Work
SiO_2	63.20	67.00
Al_2O_3	15.00	17.10
Fe_2O_3	11.00	5.60
K_2O	0.15	0.10
Na_2O	0.10	-
SO_3	0.10	0.30
LOI (Loss on Ignition)	8.80	7.60

Source: Adepegba [3]

Effect of water/cement ratio on the workability of laterized concrete mixes
The variations of slump with water cement ratio for the concrete mixes with varying laterite contents are presented in Figs. 1-3. For all the mix ratios investigated and for each laterite content, the workability of the mixes increases with increasing water/cement ratio and decreases with increasing lateritic content for each water cement ratio investigated. Higher water content requirement of mix having high laterite content does not necessarily increase the workability of the mix; it seems a good portion of the mixing water was absorbed by the laterite fines rather than lubricate the mix. A larger percentage of the laterized concrete mixes (especially having laterite

content greater than 25%) fall within the low workability range. This might call for mechanical compaction when used in concrete construction.

Effect of water/cement ratio on compressive strength of laterized concrete
Figures 4 to 6 show the variation of the 28-day compressive strength of the cube specimens with water/cement ratio for the three mixes investigated. For all the mix ratios and at all percentage levels of laterite contents (except for mix ratio 1:1:2 at 75% and 100% laterite content), the compressive strength of the laterized concrete cube specimens decreases with increasing water/cement ratio, the change becoming less pronounced at higher water cement ratios. Figure 4 reveals that at w/c ratios of 0.55 and 0.60, laterized concrete mixes 25%, 50% and 75% laterite contents gave fairly comparable compressive strengths and with average values of 31.4MPa and 28MPa respectively. For mix ratio 1:1:2 at 75% and 100% laterite levels, the compressive strength of the laterized concrete cube specimens reaches an optimum value at a w/c of 0.55 and thereafter decreases. For mix ratio 1:1:2 at 25% and 50% laterite contents, the recorded values of the compressive strength of the laterized cube specimens at a w/c of 0.60 are 28.7 and 29.4MPa respectively, representing just a 2.4% increase. However, at lower water/cement ratios of 0.40, 0.45, 0.50 and 0.55, the percentage decrease in compressive strength as laterite content increases from 25% to 50% are 9.8%, 10.8%, 6.9% and 4.5% respectively.

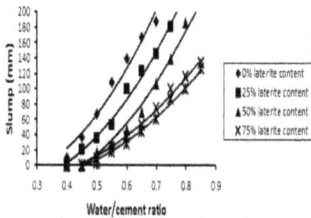

Figure 1: Variation of slump with water/cement ratio (mix ratio 1:1:2)

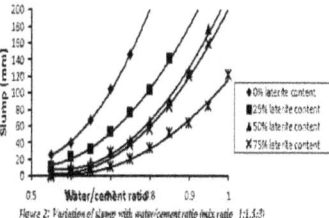

Figure 2: Variation of slump with water/cement ratio (mix ratio 1:1.5:3)

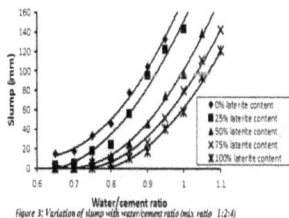

Figure 3: Variation of slump with water/cement ratio (mix ratio 1:2:4)

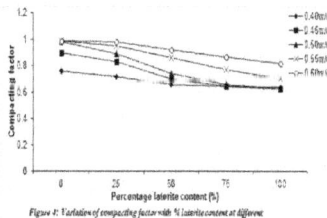

Figure 4: Variation of compacting factor with % laterite content at different water/cement ratios for mix ratio 1:1:2

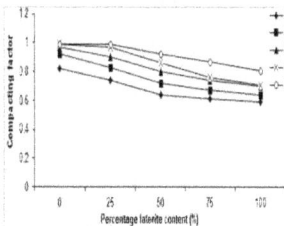

Figure 5: Variation of compacting factor with % laterite content at different water/cement ratios for mix ratio 1:1.5:3

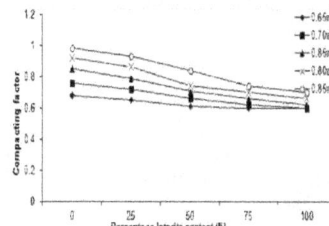

Figure 6: Variation of compacting factor with % laterite content at different water/cement ratios for mix ratio 1:2:4

Advances in Cement and Concrete   Materials Research Forum LLC
Materials Research Proceedings 51 (2025) 138-145   https://doi.org/10.21741/9781644903537-16

Thus, as w/c ratio increases from 0.4 to 0.55, the compressive strength decreases at a decreasing rate when the laterite content increases from 25% to 50%. The increasing percentage laterite content appears to have greater effect than the increase in water/cement ratio within this range. The trend observed above is reversed at a water cement ratio greater than or equal to 0.60. At 25% laterite content, the compressive strength of tested cube specimens decreases from 32.7MPa to 28.7MPa when w/c increases from 0.55 to 0.60. This represents a 12.2% decrease and can be attributed to excess free water present in the mix. At 50% laterite content and the same water cement ratio of 0.60, the quantity of free water present is reduced with a certain percentage of this absorbed by the increase in laterite content. Hence, a 2.4% improvement in compressive strength results as earlier highlighted. For the same reason, a lesser percentage decrease (6.1% compared with 12.2% at 25% laterite content) in compressive strength was recorded at 50% laterite content when w/c ratio was increased from 0.55 to 0.60. Inferring from the test results, it appears w/c ratios of 0.55 and 0.60 would be most suitable for laterized concrete mix, 1:1:2 having 25% to 75% laterite content if strength and workability considerations are combined. Similarly, w/c ratios of 0.60 and 0.65 will be adequate for the same mix ratio having 100% laterite comment. Similar explanations abound for the trend observed in Figure 5 for mix ratio 1:1.5:3.

For mix ratio 1:2:4 and as shown in Fig. 6, it appears that there is no appreciable difference between the recorded compressive strength at w/c ratios greater than or equal to 0.70 for 50% and 75% laterite contents. Based on considerations bordering on workability and compressive strength, Table 4 lists out w/c ratios suggested for various laterized concrete mixes.

Table 4 *Suggested water/cement ratios for laterized concrete mixes (and maximum corresponding 28-day compressive strength in MPa of laterized concrete given in brackets)*

Mix Ratio	Percentage Laterite Content			
	Between 0% and 25%	Between 25% and 50%	Between 50% and 75%	Between 75% and 100%
1:1:2	0% 0.45 – 0.50 (40)			
	Others 0.50-0.60 (30)	0.55 – 0.65 (25)	0.55 – 0.70 (25)	0.60 – 0.75 (20)
1:1.5:3	0% 0.55 – 0.65 (25)			
	Others 0.60-0.70 (25)	0.70 – 0.80 (15)	0.70 – 0.80 (15)	0.75 – 0.85 (10)
1:2:4	0% 0.70 – 0.80 (20)			
	Others 0.80-0.90 (15)	0.80 – 0.90 (15)	0.85 – 0.95 (15)	0.85 – 1.00 (10)

Effect of delay in casting on workability of fresh laterized concrete mix
Figures 7 to 9 show the effect of delay in casting on the workability tests. In each case, the workability of the laterized concrete mixes decreases as the time duration of delay in casting increases. This could be attributed to loss of water due to evaporation and continuation of water absorption by the aggregates. After 30 minutes, the measured slump reduces at a decreasing rate for laterized concrete mixes 1:1:2 (0% to 100% laterite contents), 1:1.5:3 (0% to 75% laterite contents) and 1:2:4 (0% to 50% laterite contents). For all other mixes, a similar observation occurs at time periods ranging between 10 to 20 minutes.

Effect of delay in casting on compressive strength of laterized concrete
Figures 10 to 12 show the effect of delay in casting on the compressive strength of laterized concrete. These reveal that there exists an optimum time when compressive strength of laterized concrete is maximum after which it begins to drop. The length of time varies for the different mixes - 40minutes for mix ratios 1:1:2 (0% to 50% laterite contents) with a gain of 28- day compressive strength of 13, 15 and 16% respectively above that of 5 minutes delay and for 1:1.5:3 and 1:2:4 both at 0% laterite content with a gain in strength of 20% and 19% respectively above that of 5 minutes delay; 30 minutes for mix ratios 1:1:2 (75% to 100% laterite contents and 1:1.5:3

Advances in Cement and Concrete

Materials Research Forum LLC

Materials Research Proceedings 51 (2025) 138-145

https://doi.org/10.21741/9781644903537-16

(25% to 75% laterite contents) and the respective gains in strength are 31, 9, 21, 21 and 25%. A range of 15 to 25 minutes was obtained for the other mixes with gain in strength above that of 5minute delay ranging between 14 to 27%. These test results indicate that beneficial effects could be achieved from delay of casting if the delay is not permitted to exceed 40 minutes for the very rich laterized concrete mixes (aggregate/cement ratio not greater than 3 and laterite content not more than 50%), 30 minutes for very rich mixes having laterite content greater than or equal to 75% and for other rich mixes with aggregate/cement ratio not greater than 5 and 15 minutes for the less rich laterized concrete mixes.

The average loss in workability within these time periods for the three cases defined above is 12mm. The reduction in slump, however, ranges between 5 to 17mm which may be assumed reasonable from the engineering standpoint.

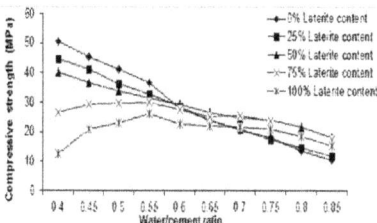

Figure 7: Effect of water/cement ratio on the 28-day compressive strength of laterized concrete (mix ratio 1:1:2)

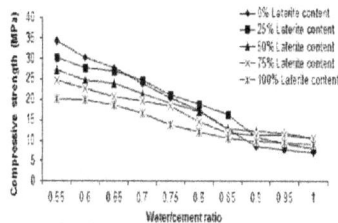

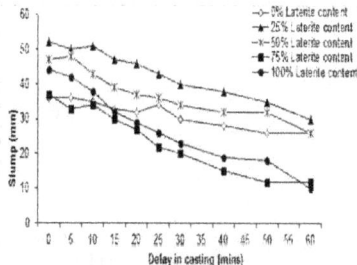

Figure 8: Effect of water/cement ratio on the 28-day compressive strength of laterized concrete (mix ratio 1:1.5:3)

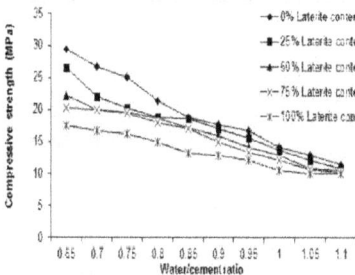

Figure 9: Effect of water/cement ratio on the 28-day compressive strength of laterized concrete (mix ratio 1:2:4)

Figure 10: Effect of delay in casting on workability of fresh laterized concrete mix (1:1:2)

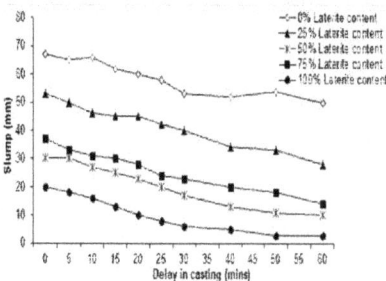

Figure 11: Effect of delay in casting on workability of fresh laterized concrete mix (1:1.5:3)

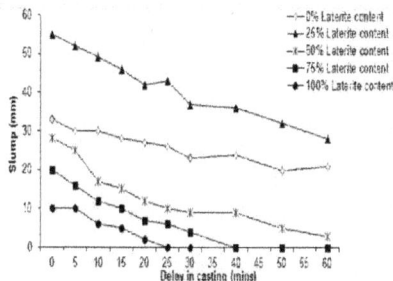

Figure 12: Effect of delay in casting on workability of fresh laterized concrete (mix ratio 1:2:4)

Conclusion and Recommendation

Conclusion

The main conclusions derived from this investigation are as follows:

- The workability of the laterized concrete mixes increases with increasing water cement ratio and decreases with increasing laterite content for each water/cement ratio investigated. It also decreases as the period of delay of casting increases. After 30 minutes of delay, the measured slump reduces at a decreasing rate for the very rich and rich laterized concrete mixes.
- The compressive strength of the laterized concrete cube specimens decreases with increasing water/cement ratio, the change becoming less pronounced at high water/cement ratio.
- Delaying the casting up to 40 minutes for the very rich mix 1:1:2 (0% to 50% laterite contents), up to 30 minutes for the rich mixes 1:1.5:3 and between 15 to 25 minutes for the less rich mix 1:2:4 had beneficial effects on the strength of laterized concrete. Averagely, a gain of about 20% was obtained in all cases with no appreciable loss in workability.

Recommendation

- Inferring from test results, adoption of water/cement ratios for laterized concrete mixes should be based on considerations of workability and compressive strength rather than on the latter alone.
- The use of superplasticizers is recommended to pave way for the use of a low water/cement ratio, yet achieving sufficient workability and higher compressive strength

References

[1] K.O, Olusola, O.O. Aina, and O., Ata, An Appraisal of the suitability of laterite for urban housing in Nigeria, Proceedings of a National Conference on 'The City in Nigeria' Ile-Ife, Niger, 9th -11th October, (2002) 168-173.

[2] C., Arum, S. A., Alabi, & R., Chinwuba, Strength and durability assessment of laterized concrete made with recycled aggregates: A performance index approach, Research on Engineering Structures and Materials 9, no. 1 (2023): 209-227.

[3] D. A., Adepegba, Comparative study of normal concrete with concrete which contained laterite instead of sand, Building Science 10, no. 2 (1975): 135-141. https://doi.org/10.1016/0007-3628(75)90029-8

[4] D., Adepegba. The effect of water content on the compressive strength of laterized concrete, Journal of Testing and Evaluation, JTEVA, 3(6) 1975 449-453. https://doi.org/10.1520/JTE11701J. https://doi.org/10.1520/JTE11701J

[5] F., Lasisi, and A. M., Ogunjide, Effect of grain size on the strength characteristics of cement-stabilized laterite soils, Building and Environment 19, no. 1 (1984): 49 - 54. doi: 10. 1016/0360-1323(84)90013-1 https://doi.org/10.1016/0360-1323(84)90013-1

[6] I., Falade, Influence of method and duration of curing and mix proportion on strength of concrete containing laterite fine aggregate. Building and Environment 26, no. 4 (1991): 453-458. https://doi.org/10.1016/0360-1323(91)90071-I https://doi.org/10.1016/0360-1323(91)90071-I

[7] K. O., Olusola, Factors affecting compressive strength and elastic properties of laterized concrete. Unpublished PhD thesis. Department of Building. Obafemi Awolowo University, Ile-Ife, Nigeria, 2005.

[8] S. B., Gowda, C., Rajasekaran, & S. C., Yaragal, Significance of processing laterite on strength characteristics of laterized concrete. In IOP Conference Series: Materials Science and Engineering vol. 431, no. 8, p. 082003. IOP Publishing. doi:10.1088/1757-899X/431/8/082003 https://doi.org/10.1088/1757-899X/431/8/082003

Materials Research Forum LLC
https://doi.org/10.21741/9781644903537-16

[9] I., Garba, J. M., Kaura, T. A., Sulaiman, I., Aliyu, & M., Abdullahi, Effects of laterite on strength and durability of reinforced concrete as partial replacement of fine aggregate. Fudma journal of sciences, 8.1, (2024), 201-207. https://doi.org/10.33003/fjs-2024-0801-2210 https://doi.org/10.33003/fjs-2024-0801-2210

[10] J. O., Ukpata, D. E., Ewa, N. G., Success, G. U., Alaneme, O. N., Otu, & B. C., Olaiya, Effects of aggregate sizes on the performance of laterized concrete. Scientific Reports, 14(1), (2024): 448. https://doi.org/10.1038/s41598-023-50998-1

[11] British Standard Institution, " Composition, specification and conformity criteria for common cements", BS EN 197: Part 1. London.British Standard Institution (BSI), 2001.

[12] British Standard Institution, "Testing fresh concrete" BS EN 12350: Parts 1-6, BSI, London, 2000.

[13] British Standard Institution, "Testing hardened concrete" BS EN 12390: Parts 1-8, BSI, London, 2000.

[14] M.A., Amjad and S.H., Alsayed, Properties of mortar, bricks, and masonry incorporating red and white sands. The Arabian Journal for Science and Engineering, Vol. 24, (1999), 169-183. https://search.emarefa.net/detail/BIM-389687

[15] O.A., Kayyali, Effect of certain mixing and placing practices in hot weather on the strength of concrete. Building and Environment 19, no.1 (1984), 59 - 63. https://doi.org/10.1016/0360-1323(84)90015-5

Advances in Cement and Concrete
Materials Research Proceedings 51 (2025) 146-155

Materials Research Forum LLC
https://doi.org/10.21741/9781644903537-17

The use of waste materials as a cement blend for developing concrete

J. SULE[1,a*], O.A. EJEMBI[2,b], A.O. SHAIBU[3,a], S.K. BONIFACE[3,c]

[1]Building Research Department, Nigerian Building and Road Research Institute, Jabi -Abuja, Nigeria

[2]Architecture Department, Ahmadu Bello University, Zaria, Nigeria

[3]Civil Engineering Department, Plateau State Polytechnic, Barkin Ladi, Nigeria

[a]jibrinsule2005@yahoo.com, [b]aoejembi@abu.edu.ng, [c]qingabdullah747@gmail.com, [d]Shetakboniface@gmail.com

Keywords: Waste Glass, Recycled Brick Waste, Portland Cement, Compressive Strength, Sustainable

Abstract. The use of Supplementary Cementitious Materials (SCMs) to offset a portion of cement is a promising alternative for the production of eco-friendly mortar. Several industrial by products have been used successfully as SCMs, however, waste glass and recycled brick waste has not yet achieved commercial success. The objective of this study is to develop blended concrete from local wastes; waste glass and recycled burnt brick as a consequence of the 'zero waste' objective of circular economy. Portland cement was substituted with 10-40% by mass, with finely dispersed waste glass and recycled burnt brick powder in the production of concrete at standard curing and using hand mixing. Physical properties of the local waste were investigated, X-ray Fluorescence (XRF) and X-ray Diffraction (XRD) analysis was also conducted to characterize the chemical and mineralogical composition of these wastes. The compressive strength of the developed blended concrete, was also investigated experimentally. It was evident that all hydrated materials (blended concrete samples) displayed increasing compressive strength with curing age. It further shows that glass develops steady but gradual strength. A 20% replacement of Portland cement with waste glass and recycled burnt brick have compressive strength comparable to the control and found convincing for structural application considering environment and cost. It is concluded that waste glass and recycled burnt brick can be an effective measure in sustainable development of cement blend for developing concrete.

Introduction

The use of supplementary cementitious materials (SCMs) to offset a portion of cement is a promising alternative for the production of eco-friendly concrete. Several industrial by products have been used successfully as SCMs ([1,2]. However, waste glass and recycled brick waste has not yet achieved commercial success [3,4,5]. From researches, notable findings indicate that waste glass and recycled burnt brick waste has a chemical composition and phase comparable to traditional SCMs [6, 7, 8] and could be used as supplementary cementitious material [9]. The properties that influence the pozzolanic behavior of waste glass, recycled clay brick waste and most pozzolans are fineness, chemical composition, and the pore solution present for reaction [2,4]. The pozzolanic properties of glass are notable at particle sizes below approximately 100 μm, similar to burnt brick waste and have pozzolanic reactivity at low cement replacement levels after 90 days of curing [10].

Previous studies by Asteray [11] with glass addition to cement were not conclusive, considering workability and strength of resulting concrete. For the incorporation of waste glass and burnt brick waste as supplementary cementitious material for blended cement, there are contradictory opinions. The introduction of recycled clay brick waste alters the equilibrium and opens up new

Advances in Cement and Concrete
Materials Research Proceedings 51 (2025) 146-155

Materials Research Forum LLC
https://doi.org/10.21741/9781644903537-17

possibilities and permutations. The suitability of the trio and optimum mix proportion has not yet been evaluated.

To solve raw material shortage, a culture of using local waste raw materials in the construction sector should be recommended [11], also keeping in view the advantages of limiting environmental pollution which makes locally available pozzolanic materials use essential [12].

This paper therefore presents a comprehensive study on the use of waste glass powder and recycled burnt brick powder in the field of blended cement. The focus of the research is to present additional information in the field of material recycling to explore the possible uses of these recyclable materials in construction. The assessment of different properties of constituents in the cement blend is presented. The current work concludes performance-based analysis and recommendations imperative to the use of blended cement produced from waste glass powder and recycled burnt brick. With the increase in construction activities and current high cost of cement, blended cement can be used as an alternative.

Methodology

Experimental design was employed as the research design for this study. As an innovative construction material (blended cement), a proof of viable local wastes; waste glass powder and recycled burnt brick waste were selected for development of cement blend. Accordingly, raw materials, experimental details and data analysis methods were discussed as follows.

Materials. Portland cement, finely dispersed waste glass powder, finely dispersed recycled burnt bricks powder and water were used for the development of blended cement mix and for the entire tests. The raw materials used in this study are described as follows:

Portland cement. Commercially available Portland cement with strength class 42.5N from BUA cement factory which conforms with Nigerian Industrial Standard (NIS 234:2012) was used in this research. The properties of cement used is shown in Table 1.

Table 1. Properties of Portland cement used in this study

Parameters	Specific surface [cm²/g]	Water demand [%]	Setting time [Minutes]		Soundness [mm]	Compressive strength (concrete cubes) [N/mm²]	
			Initial	Final		At 3 Days	At 28 days
Results	2,396	25.65	240	380	0.3	18.49	40.93

Fine aggregate. Locally available river sand from Jere, Kaduna State was used as fine aggregate in the preparation of all test specimens. Fine aggregates were prepared according to graded sand requirement based on British Standard (BS 882, 1992 BS812-103.1, 1995) and were used in dry condition. The grading of fine sand used for the entire mixtures is shown in Table 2.

Table 2. Grading of fine aggregate used in this study

Sieve size [mm]	Mass retained [g]	% Retained	% Passing
4.75	5.0	1.0	99.0
2.00	57.0	6.0	93.0
0.60	426.0	47.0	46.0
0.212	382.0	42.0	4.0
0.063	34.0	4.0	0.0

Figure 1 shows the particle size distribution for fine aggregates while the physical properties of fine sand is shown in Table 3.

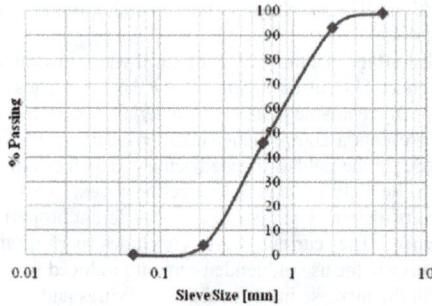

Fig 1. Particle Size distribution for fine aggregates

Table 3. Physical properties of fine aggregate used

Parameters	Results
Specific gravity	2.55
Loose density [g/cm^3]	1571.60
Bulk density [g/cm^3]	1747.00
Fineness modulus	2.81

Coarse aggregate. Two sets of aggregate were obtained from Dutse quarry site in Abuja, based on grading requirement for M25 concrete. They are 20 mm and 10 mm aggregates respectively.

Water. Portable drinking water, free from organic solid materials conforming to NIS 456:2000 was used for the concrete mix with pH value between 7.5- 8.5.

Finely dispersed local waste powders. Glass cullet obtained from construction sites was first transformed into powder by milling in a laboratory ball mill and passed through a 90 μm sieve. Only glass powder with size lower than 90 μm was used. 90 μm particle size is used as the standard requirement for cement production ASTM, C3124 (ASTM, 2017). Research shows that glass powder exhibits pozzolanic reactivity at 90 μm particle size and below, thus 90 μm particle size was adopted. Clay brick wastes were sourced locally from River side brick production site in Makurdi, Benue state. It was crushed, milled and sieved, grains passing through 90 μm sieve was the primary material used. Figure 2 shows the preparation of finely dispersed local waste powders for this study.

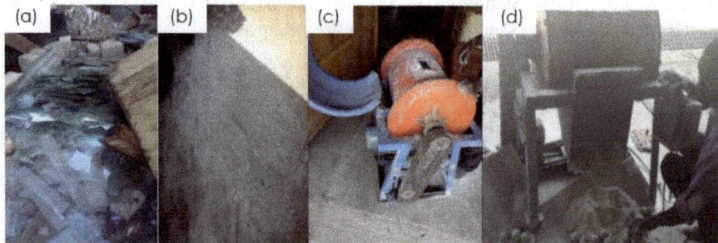

Fig 2. Preparation of finely dispersed local waste powders (a) waste glass (b)mill glass powder (c) mill ball (d) brick powder

Advances in Cement and Concrete

Materials Research Proceedings 51 (2025) 146-155

Materials Research Forum LLC

https://doi.org/10.21741/9781644903537-17

Physical properties of waste glass powder and recycled burnt brick used are shown in Table 4.

Table 4. Physical properties of local wastes used

Parameters	Results	
	Waste glass powder	Recycled burnt brick powder
Specific gravity	2.56	2.79
Loose density [g/cm^3]	2505.12	2446.40
Bulk density [g/cm^3]	2555.04	1747.00
Fineness modulus	3.00	3.73

Experimental Design

Proposed mix design for Blended concrete. For this study, Concrete was developed based on the existing mix proportions for production of concrete, mass ratio of cement to fine aggregate is set at 1:3 [13] and in conformity with NIS standard. Since the mix design was employed using local waste materials (waste glass and recycled burnt brick) together with Portland cement as core raw materials, the name "Blended Concrete" was given for the concrete mixture. In general, Portland cement was replaced partially by combination of paired local wastes with different percentages from 0% to 40% by weight. Table 5 shows the mixed design for blended cement used.

The control mix was developed from Portland cement, fine sand and water. For all mixes, hand mixing, standard water curing and uniform water-binder ratio of 0.50 were used. As shown in Table 5, Twenty-seven mixes were utilized including the control mix, though sixteen gives a combination of blended concrete. Raw material characterization. In this study, the physical properties of finely dispersed local wastes such as specific gravity, fineness module, Bulk density and specific gravity were investigated. In addition, the X-ray fluorescence (XRF) spectrometry analysis was also conducted. Mixed proportion, specimen preparation and curing procedure. The mix propositions designed in Table 5 was applied.

Table 5. Mixed design for blended cement used

Specimen details	Cement [%]	Glass Powder [%]	Fire Bricks [%]	Sand [g]	Water (w/c)
Control sample R (100% cement)	100	0	0	480	0.50
90% cement (at 10% replacement)	90	10	0	480	0.50
	90	5	5		
	90	0	10		
80% cement (at 20% replacement)	80	20	0	480	0.50
	80	15	5		
	80	10	10		
	80	5	15		
	80	0	20		
70% cement (at 30% replacement)	70	30	0	480	0.50
	70	25	5		
	70	20	10		
	70	15	15		

	70	10	20	
	70	5	25	
	70	0	30	
60% cement (at 40% replacement)	60	40	0	480	0.50
	60	35	5	
	60	30	10	
	60	25	15	
	60	20	20	
	60	15	25	
	60	10	30	
	60	5	35	
	60	0	40	

Blended concrete specimens were prepared based on British Standard by preparing watertight and non-absorbent $50 \times 50 \times 50$ mm³ cube moulds (BS 5328: Part 1, 1997). In order to prepare specimens, dry mixing of ingredients was done, followed by wet mixing until a visually acceptable mix was obtained. After all the preparation, the standard curing of the test specimens was done for 3, 7, 28 and 90 days in water at a temperature of 20 ± 2 °C.

Data analysis. The results of the chemical analysis for local waste materials proposed in this study were compared using graphical descriptions with the minimum requirement for a standard pozzolana quality as per ASTM C618. In addition, the results were analyzed statically using Microsoft Excel through tables and graphs that show the most relevant properties related to the study.

Results and Discussion
Material characterization
In this study, specific gravity, fineness module and density for each waste raw material was done. Additionally, the overall chemical compositions of the wastes as pozzolans were evaluated. Table 6 shows the chemical compositions of local waste raw materials used.

Table 6. Chemical composition of local wastes used

Element/Compound	OPC (43.5N)	Local wastes Powder	
	Cement [%]	Glass [%]	Fired bricks [%]
SiO_2	18.9	80.77	53.99
Al_2O_3	3.13	1.35	19.83
Fe_2O_3	1.14	0.97	12.11
CaO	65.23	12.83	0.96
MgO	1.32	3.50	4.60
K_2O	0.75	0.19	2.55
Na_2O	5.10	0.39	0.68
TiO_2	0.05	0.08	1.17
SO_3	-	0.24	0.12
MnO	0.02	0.05	0.73
P_2O_5	0.12	0.19	0.35
LOI	1.33	0.33	2.33

Advances in Cement and Concrete
Materials Research Proceedings 51 (2025) 146-155

Materials Research Forum LLC
https://doi.org/10.21741/9781644903537-17

The overall chemical composition of a pozzolan is considered as one of the parameters governing long-term performance (e.g. compressive strength) of the blended cement binder, ASTM C618 prescribes that a pozzolan should contain $SiO_2 + Al_2O_3 + Fe_2O_3 > 70$ wt.%.

Table 7 shows the major chemical compositions (silica, alumina and iron oxide content) of the local wastes in this study after XRF analysis. As evaluation criteria, the sums of the three mineral components (silica, alumina and iron oxide content) in each local waste were compared with the requirement of ASTM C618 as a pozzolan. In view of that, finely dispersed glass powder contains 83.09% and finely dispersed recycled burnt brick powder 85.93% by weight. The chemical compositions for the proposed raw materials were greater than 70 (by weight %). Hence, they fulfil the requirement of ASTM C618 as a pozzolan.

The identified crystalline compounds were quartz, feldspars (mainly as anorthite), and low content of hematite for waste glass powder; and quartz, feldspars (albite and microcline), and mica (phlogopite) for recycled burnt brick powder.

The mineralogical composition of both binder constituents was determined by XRD analysis using Philips PW 3710 diffractometer (Phillips, Netherland) at Umaru Musa Yar'adua University Katsina. XRD patterns of studied powders are shown in Figure 3 and Figure 4.

Table 7. Chemical composition of local wastes used

Minerals	Finely dispersed glass powder [%]	Finely dispersed glass powder [%]
Silica (SiO_2)	80.77	53.99
Alumina (Al_2O_3)	1.35	19.83
Iron oxide (Fe_2O_3)	0.97	12.11
Total	83.09	85.93

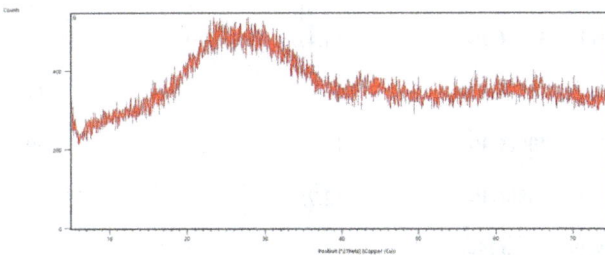

Fig 3. X-ray diffraction pattern of waste glass powder

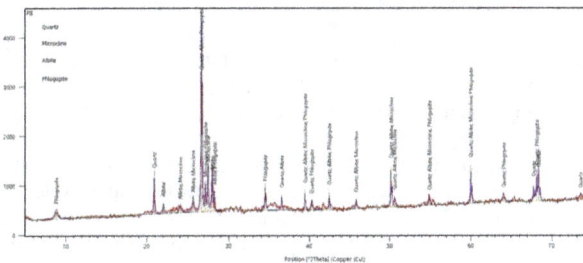

Fig 4. X-ray diffraction pattern of burnt bricks used

The effect of waste glass powder and recycled burnt brick on mechanical properties of blended cement. In this study, compressive strength was evaluated by replacing different percentages of the proposed materials. The mean compressive strengths of four blended cement specimens produced from sixteen categories (including control mix) containing Portland cement, glass powder and recycled burnt brick powder is presented in Table 8 and Figure 5.

The compressive strength test method is in conformity with ASTM C 109 (2016). As shown in Table 8, the compressive strength increases with the curing age in all mix. In early age at 3 days standard curing, mix 1 and mix 4 developed strengths of 24.67 and 20.9 MPa respectively greater than the Control specimen that developed strength of 18.49 MPa. Mix 2 and Mix 3 exhibited strengths of 18.32 and 18.11 MPa, accordingly.

Table 8. Mechanical properties of blended cement containing Portland cement (PC), glass powder (GP) and recycled burnt brick powder (BP)

Mix designation	Compressive strength [MPa]			
	3 Days	7 Days	28 days	90 Days
Control	18.49	28.43	40.93	45.2
Mix 1 (90%PC-5%GP-5%BP)	24.67	26.26	36.48	39.3
Mix 2 (80%PC-15%GP-5%BP)	18.32	25.21	31.32	36.8
Mix 3 (80%PC-10%GP-10%BP)	18.11	29.48	35.4	40.1
Mix 4 (80%PC-5%GP-15%BP)	20.9	31.68	36.52	41.95
Mix 5 (70%PC-25%GP-5%BP)	15.15	20.8	26.13	31.9
Mix 6 (70%PC-20%GP-10%BP)	16.24	20.81	28.27	29.1
Mix 7 (70%PC-15%GP-15%BP)	13.54	19.05	29.36	30.4
Mix 8 (70%PC-10%GP-20%BP)	15.3	20.43	29.24	34.6
Mix 9 (70%PC-5%GP-25%BP)	16.1	19.61	27.06	28.2
Mix 10 (60%PC-35%GP-5%BP)	11.91	15.84	24.52	17.3
Mix 11 (60%PC-30%GP-10%BP)	14.47	20.78	28.205	27.95
Mix 12 (60%PC-25%GP-15%BP)	15.75	19.79	29.47	29
Mix 13 (60%PC-20%GP-20%BP)	13.58	20.54	26.99	31.45
Mix 14 (60%PC-15%GP-25%BP)	12.75	20.93	29.45	30.7
Mix 15 (60%PC-10%GP-30%BP)	12.14	20.87	26.2	22.7
Mix 16 (60%PC-5%GP-35%BP)	11.71	17.56	22.83	23.8

At age 7days standard curing, mix 3 and mix 4 exhibited strengths of 29.48 and 31.68 greater than the control with strength of 28.43 MPa. However, at 28 days standard curing, the control mix exhibited the highest strength of 40.93 MPa followed by Mix 4, Mix 1 and Mix 3 with strength of 36.52, 36.48 and 35.4 MPa, accordingly.

At age 90 days standard curing, Mix 4 had developed strength of 41.95 MPa , Mix 3; 40.1 MPa and mix 1; 39. 3 MPa. The effect of waste glass powder and recycled burnt brick on the microstructure properties of blended cement. In this study, Scanning Electron Microscope analysis (SEM) was used as the microstructure investigation technique. The SEM analysis has been made over all the samples, considering ageing time coupled with materials composition, but for brevity,

Advances in Cement and Concrete
Materials Research Proceedings 51 (2025) 146-155

Materials Research Forum LLC
https://doi.org/10.21741/9781644903537-17

only the most representative images for high performing blended cement and the reference (control) have been reported.

The microstructure of the reference concrete (control) as shown in Figure 5a, appears homogeneous and dense. This, together with the high mechanical strength of river sand and the good adhesion observed between sand and cementitious paste, explains the high mechanical performance of the reference concrete. In the case of Mix 4 concrete (Figure 5b), the microstructure appears quite homogeneous and dense, however, the Interfacial Transition Zone (ITZ) between cement paste and waste glass and recycled brick aggregate is detectable only with difficulty, meaning there is a good adhesion between aggregate and binder paste. This is probably due to the possible (at least superficially) pozzolanic reactivity of this unconventional binder with Ca(OH)$_2$ of the cement paste, as already reported by other authors [14].

This justifies the high mechanical behaviors of this concrete in comparison to the conventional sand concrete, in spite of the higher lightness. When Mix 9 and 12 blend was used (Figure 5c and 5d), the microstructure of concrete appears very heterogeneous and porous. The ITZ between pozzolans and cement paste is easily detectable meaning a poor adherence between the two components. It is observed that there are large waste glass and recycled pozzolan particle (dark) strictly entrapped by the cementitious matrix; the interface appears well defined with no voids. These facts explain why Mix 12 concrete shows poor mechanical performance. The ITZ between Mix 4 pozzolan and cement paste is good. This explains why its mechanical strength is significantly higher compared to that of Mix 9 and 12 concretes, even if still lower than that of the reference concrete.

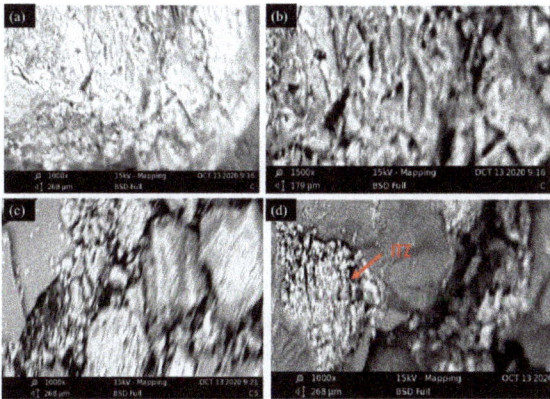

Fig 5. *SEM images of blended concrete: (a) Control Concrete (b) Mix 4 (80%Pc-5%GP-15%BP) (c) Mix 9 (70%PC-5%GP-25%BP) (d) Mix 12 (60%PC-25%GP-15%BP)*

Conclusion
The production of stable concretes using BUA Portland cement, waste glass powder, recycled burnt bricks, fine aggregate and water were investigated. Concretes were produced using a fixed cement/ aggregate ratio (1/3), whereas cement, waste glass and recycled burnt brick was added in different proportions. Based on the result the following important conclusions were drawn:
 i. The finely dispersed glass powder contains 83.09% and finely dispersed burnt brick powder contain 85.93% by weight as per the requirement of ASTM C618 as a pozzolan.
 ii. All hydrated materials displayed increasing compression strength with curing age from 3, 7, 28, and 90 days of curing in a moist environment.

iii.   Glass containing samples showed a rapid increase of strength with respect to the reference compositions when subjected to long-term ageing.

iv.   In mix designs from 10 -30% replacement of cement, good pozzolanic effect and good density were observed.

v.   Mix 3 (80%PC-10%GP-10%BP) and Mix 4 (80%PC-5%GP-15%BP) showed impressive compressive strength and good microstructure

Acknowledgements

The authors are thankful to Nigerian Building and Road Research Institute for the financial support and laboratory facilities; Umaru Musa Yar'adua University, Katsina for the laboratory facilities in this study.

References

[1] Islam, G., Islam, M., Akter, A., Islam, M, Green Construction Materials-Bangladesh perspective. In. Proceedings of the International Conference on Mechanical Engineering and Renewable Energy (ICMER2011). Chittagong, Bangladesh. (2011).

[2] Imbabi, M., Carrigan, C., Mckenna, S, Trends and developments in green cement and concrete technology. International journal of Sustainable Built Environment, 1 (2013) 194-216. doi:10.1016/j.ijsbe.2013.05.001 https://doi.org/10.1016/j.ijsbe.2013.05.001

[3] Islam, G., Rahman, M., Kazi, N, Waste glass powder as partial replacement of cement for sustainable concrete practice. International Journal of Sustainable Built Environment, 6 (2017), 37-44. doi:10.1016/j.ijsbe.2016.10.005 https://doi.org/10.1016/j.ijsbe.2016.10.005

[4] Rasheed, A, Recycled waste glass as fine aggregate replacement in cementitious materials based on Portland cement. Constr. Build. Mater, 72 (2014), 340-357. https://doi.org/10.1016/j.conbuildmat.2014.08.092

[5] Apebo, N., Agunwamba, J., Ezeokonkwo, J, Suitability of crushed over burnt bricks as coarse aggregates for concrete. International Journal of Engineering Science and Innovative Technology (IJESIT), 3 (2014) 316-321.

[6] Binici, H., Aksogan, O., Cagatay, I., Tokyay, M., Emsen, E, The effect of particle size distribution on the properties of blended cements incorporating GGBFS and natural pozzolan (NP). Powder Technol., 177 (2007) 140-147. https://doi.org/10.1016/j.powtec.2007.03.033

[7] Bhattacharjee, E., Nag, D., Sarkar, P., Haldar, L. An Experimental Investigation of Properties of Crushed Overburnt Brick as Aggregates Concrete. International Journal of Engineering Research and Technology, 4 (2011) 21-30.

[8] Nassar, R., Soroushian, P, Strength and durability of recycled aggregate concrete containing milled glass as partial replacement for cement. Constr. Build. Mater., 29 (2012). 368-377. https://doi.org/10.1016/j.conbuildmat.2011.10.061

[9] Ling, I., Teo, D, EPS RHA concrete bricks - A new building material. Jordan Journal of Civil Engineering, 7 (2013) 361-370.

[10] Shi, C., Riefler, C., Wang, H, Characteristics and pozzolanic reactivity of glass powders. Cem. Concr Res, 35 (2005) 987-993. https://doi.org/10.1016/j.cemconres.2004.05.015

[11] Asteray, B., Oyawa, W., Shitote, S, Experimental investigation on compressive strength of recycled reactive powder concrete containing glass powder and rice husk ash. Journal of Civil Engineering Research, 7 (2017) 124-129. doi:10.5923/j.jce.20170704.03

[12] Khan, A., Khan, B, Effect of partial replacement of cement by mixture of glass powder and silica fume upon concrete strength. International journal of Engineering Works, 4 (2017) 124-135.

[13] Stefano, M., Gabriele, T., Erika, F, Recycling Glass Cullet from Waste CRTs for the Production of High Strength Concretes. Journal of Waste Management. (2013) 63-71. doi:10.1155/2013/102519 https://doi.org/10.1155/2013/102519

[14] Pavli'k, V., Uza'kova, M, Effect of curing conditions on the properties of lime, lime-metakaolin and lime-zeolite concretes. Construction and Building Materials, 102 (2016) 14-25. https://doi.org/10.1016/j.conbuildmat.2015.10.128

Advances in Cement and Concrete
Materials Research Proceedings 51 (2025) 156-166

Materials Research Forum LLC
https://doi.org/10.21741/9781644903537-18

Consistency and strength indices of geopolymerized crude oil-contaminated soil using alkaline activated blast furnace slag

Johnson R. OLUREMI[1,a], Olalere K. SULAIMAN[1,b*] and Walied A.H. ELSAIGH[2,c]

[1]Department of Civil Engineering, Ladoke Akintola University of Technology, Ogbomoso, Nigeria

[2]Department of Civil Engineering, University of South Africa, Florida Science Campus, Johannesburg, South Africa

[a]jroluremi@lautech.edu.ng, [b]olladeem2@gmail.com, [c]hussiwam1@unisa.ac.za

Keywords: Activator, British Standard (BS), California Bearing Ratio, Crude Oil Contaminated Soil (COCS), Ground Granulated Blast Furnace Slag (GGBFS), West African Standard (WAS)

Abstract. Crude oil contamination has been proven to adversely alter soil's geotechnical properties, which normally renders it unsuitable for use as road construction materials. Although researchers have extensively studied various ways of remediating and improving the properties of crude oil-contaminated soils (COCS), the differing results still make it a subject of interest. This study investigated the performance evaluation of the effects of ground granulated blast furnace slag (GGBFS) activated with a mixture of Sodium silicate and Sodium hydroxide on the strength of crude oil-contaminated soil (COCS) at different percentages by weight for safe reuse in highway construction applications. Particle size distribution, Atterberg limits, strength characteristics and microstructural and elemental study of crude oil contaminated soil obtained from Baranyowa Dere village of Ogoni land, Gokana Local Government Area, River State, Nigeria, and admixed with alkaline activated slag was studied. The interaction between COCS - GGBFS – Activator (NaOH+Na$_2$SiO$_3$) led to the reduction in the Optimum Moisture Content (OMC) increase in the Maximum Dry Density (MDD) of the samples. The California Bearing Ratio (CBR) also increased for soaked and unsoaked conditions, respectively. The percentage of silica in the admixtures reduced with increasing percentage of the additives. It was concluded that the use of alkaline-activated GGBFS is a promising stabilizer for improving the COCS's engineering properties for use in highway pavement construction.

Introduction

There are several ways that crude oil or petroleum products might pollute soil, viz pipeline leaks and vandalism, ship accidents, onshore and offshore crude oil exploration, discharge from coastal facilities, natural seepage, underground storage facilities leakage and onsite oil spillage [1]. This contamination affects the quality of the original soil and alters its engineering properties. The potential reuse of contaminated soil as construction materials has been seen as one of the best alternative means of its disposal; however, since soil to be recycled or reused must be classified as non-hazardous according to the New Jersey Department of Environmental Protection [2], it needs to be treated to ensure its decontamination to the acceptable level [1]. For a very long time, the standard stabilizers used in soil stabilization applications were cement and lime. The extensive production and application of these two chemicals among others have caused serious impacts on the environment and health hazards. However, due to the high carbon emission in cement production, research on soil stabilization using more environmentally friendly binders with lower carbon footprints has attracted much attention in recent years.

Geopolymers, known as the new generation of green materials for cement substitution, are synthesized from the chemical reaction between an amorphous precursor, which is rich in alumina

Advances in Cement and Concrete
Materials Research Proceedings 51 (2025) 156-166

Materials Research Forum LLC
https://doi.org/10.21741/9781644903537-18

(Al_2O_3) and silicate (S_iO) with a sodium or potassium-based activator [3]. Geopolymer offers a better alternative to Portland cement, with its high strength, low cost, low energy consumption, and CO_2 emissions during synthesis. The most widely used activators for the activation of precursors in geopolymers are sodium hydroxide (NaOH), sodium silicate (Na_2SiO_3), or a mixture of them [4]. In soil stabilization, the addition of an activated geopolymer as an alternative binder on cement will help combat the problem of CO_2 emission and energy consumption while maintaining the good engineering properties of the soil.

This research work examines the strength characteristics and environmental performance of alkaline activated slag stabilized crude oil contaminated soil from Baranyowa Dere village of Ogoni land, Gokana Local Government Area, River state, Nigeria. During this research, the strength evaluation tests: Compaction and California Bearing Ratio (CBR) were carried out. Also, microstructural investigations of the optimal GGBFS stabilized COCS were examined.

Materials and Methods

Materials
The materials used for this study are Crude Oil-Contaminated Soil (COCS), Blast furnace Slag (BFS), Sodium Hydroxide, Sodium Silicate and water.

Sample Collection
The Crude oil-contaminated soil used was collected from an oil-contaminated borrow pit (Latitude 4.653547°N and Longitude 7.27275°E) at a depth of 1.0m in Baranyowa Dere village of Ogoni land, Gokana Local Government Area, River State, Nigeria. The raw Blast Furnace Slag (BFS) was collected from African Foundries Limited, Km 45, Shagamu - Ikorodu Expressway, Ogijo, Ogun State, Nigeria. Both Sodium Hydroxide (Pellet) and Sodium Silicate (solution) were purchased at local Chemical Stores, at Ogunpa, Ibadan, Oyo State, Nigeria. Water was sourced from a nearby borehole around the laboratory.

Characterization of materials
Chemical composition analyses of COCS and ground granulated blast furnace slag (GGBFS) were carried out at Covenant University Central Instrumentation Research Facility, Ota, Ogun State, Nigeria using X-ray Fluorescence (XRF), Spectrometer as specified in BS EN 196-2 (1995), to determine its oxides composition as pozzolanic material. XRD spectra mixture of raw crude oil-contaminated soil and granulated blast furnace slag (GGBFS) were analysed to determine the presence and amount of minerals species as well as identify crystalline phases.

The suitability of GGBFS concerning its chemical composition for alkali-activated systems was determined by the Alkalinity factor (Kb) in Eq. 1 and Hydration Modulus (HM) in Eq. 2. The utilized slag alkalinity was evaluated as the ratio between the amounts of basic and acid oxides using [6, 7]. Slags with a Kb equal to 1 were classified as neutral, while Kb > 1 slag was considered basic and Kb < 1 as acidic. The hydration modulus (HM) higher than 1.4 indicates efficient hydration during the alkali activation process [8, 9].

$$K_b = \frac{CaO + MgO + Fe_2O_3 + K_2O + Na_2O}{SiO_2 + Al_2O_3} \tag{1}$$

$$HM = \frac{CaO + MgO + Al_2O_3}{SiO_2} \tag{2}$$

Total Petroleum Hydrocarbon (TPH)
Extraction of crude oil Total Petroleum Hydrocarbon (TPH) in the crude oil-contaminated soil was determined using gravimetric analysis, following saponification in methanolic-KOH, extraction by n-Hexane, and separation via liquid chromatography.

Advances in Cement and Concrete   Materials Research Forum LLC
Materials Research Proceedings 51 (2025) 156-166   https://doi.org/10.21741/9781644903537-18

Chemical composition of silicate solution
The laboratory analysis was carried out on sodium silicate (Na_2SiO_3) solution to determine the percentage of Silica (SiO_2), Sodium Oxide (Na_2O) and water (H_2O) present.

Sample preparation
The COCS sample was air-dried and then pulverized with the aid of hands to remove the lumped soil and sieved through sieve No 4 (4.75*mm*) aperture. The BFS was ground to powder form. It was later sieved through sieve No. 200 to remove impurities and achieve fineness material known as Ground Granulated Blast Furnace Slag (GGBFS) similar to cement in fineness.

For this study, the concentration of 8 Mole of NaOH was calculated as the molecular weight of NaOH which is 40 [5]. The alkaline solutions were left for 24 hours to ensure complete dissolution before use. The NaOH solution was then added to the Na_2SiO_3 solution by volume in a ratio of 1:2 to create the activating solution for the synthesis of geopolymer for the study. Then, the GGBFS in powder form was weighed at percentages of 5, 10, 15 and 20% by the weight of the soil while the activator ranged from 1 to 4% (see Table 1). The dry soil and slag were a mixed manually and blended for ten minutes for a homogeneous mixture. The activator solution was then added to the mixture and mixed well for 10 minutes. The four (4) runs of samples with varying proportions of Soil, GGBFS, and Activator compositions were presented in Table 1. The mixing was continued until homogeneity was achieved before carrying out any geotechnical test.

Table 1: Design mix for the four samples

GGBF Slag [%]	0	5	10	15	20
Activator [%]	0	1	2	3	4
Soil [%]	100	94	88	82	76

Laboratory Experimentation
The following geotechnical tests: Sieve analysis, Liquid, Plastic and Shrinkage limits, British proctor (BC), West Africa Standard Compaction (WAS) and California Bearing Ratio (CBR) were carried out on the alkaline activated GGBFS stabilized COCS. Scanning Electron Microscopy (SEM) was used to analyse the internal structure of both control COCS and COCS stabilized with GGBFS+($NaOH+Na_2SiO_3$).

Results and Discussion

Chemical Composition of GGBFS and COCS
Chemical compositions of GGBFS and COCS are presented in Table 2. The percentage of SiO_2, Al_2O_3 and Fe_2O_3 in GGBFS and COCS are 29.34, 17.76 and 5.62%, and 80.44, 14.37 and 0.86%, respectively. The quantity of silica was higher in the COCS than in the GGBFS because the soil is sandy in nature. The alumina and ferrite were higher in percentages because the slag is a by-product of smelting of iron scraps. The Kb of GGBFS was 0.94, GGBFS was classified as slightly acidic While HM was 1.52 which is higher than 1.4 indicating that GGBFS has good hydration properties. The soil is sandy in nature because according to [10] for a soil to be sandy or other than laterite, the ratio of SiO_2/Al_2O_3 should be greater than or equal to 2.0. The ratio for the soil used was 5.57 which was higher than 2.0 indicating that the soil is not lateritic but sandy.

Advances in Cement and Concrete   Materials Research Forum LLC
Materials Research Proceedings 51 (2025) 156-166   https://doi.org/10.21741/9781644903537-18

Crystalline Phases in GGBFS and COCS

The crystalline phases existing in the raw soil are Quartz, Aibite, Orthoclase, Muscovite, and Osumilite. The presence of this crystalline species in the soil confirmed sandy nature of the soil as inferred from the result of the ratio of silica to alumina present in the soil. This will ensure good reactivity with precursors or cementitious materials. See Fig. 1a. The result of GGBFS in Fig. 1b is similar to that reported by [11, 12] which showed that the main mineralogical components are larnite, alite, mayenite, brownmillerite, gehlenite, wustite, magnetite, quartz and periclase. Also, [13], listed among others these mineralogical components Portlandite, Srebrodol'skite, Merwinite, Larnite, Calcite (manganoan), Lime, Dolomite, Wollastonite, Periclase, and Pentahydrate.

Table 2: Chemical Composition of COCS and GGBFS

Oxide name	Conc. in COCS [%]	Conc. in GGBFS [%]
Fe_2O_3	0.861	5.621
SiO_2	80.435	29.342
Al_2O_3	14.370	17.756
MgO		2.208
CaO	0.347	35.994
K_2O	0.295	0.655
SO_3	0.472	0.425
TiO_2	1.573	2.208
BaO	0.072	0.441
Cr_2O_3	0.138	0.796
MnO	0.028	2.817
P_2O_5	0.338	
Others		1.737

Phase name	Formula	Figure of merit	Phase reg. detail	Space Group	DB Card Number
Quartz, syn	Si O2	0.589	S/M(PDF-4 Minerals 2024)	154 : P3221	04-014-7569
Orthoclase	Al2 O3 · K2 O ·6 Si O2	1.787	Import(PDF-4 Minerals 2024)	12 : C12/m1	00-002-0475
Albite	Na Al Si3 O8	3.212	Import(PDF-4 Minerals 2024)	2 : C-1	00-001-0739
Muscovite	H2 K Al3 (Si O4)3	3.103	Import(PDF-4 Minerals 2024)	15 : C12/c1	00-001-1098
Osumilite	K · Na · Ca · Mg · Fe · Al · S..	3.133	Import(PDF-4 Minerals 2024)	192 : P6/mcc	00-010-0413

Phase Data View

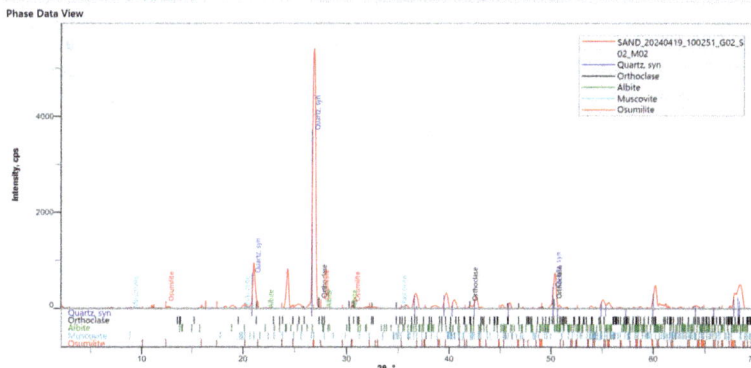

A: XRD for COCS

Phase name	Formula	Figure of merit	Phase reg. detail	Space Group	DB Card Number
Dolomite	Ca Mg (C O3)2	0.797	S/M(PDF-4 Minerals 2024)	148 : R-3H	01-075-1759
Lime, syn	Ca O	3.118	Import(PDF-4 Minerals 2024)	225 : Fm-3m	00-004-0777
quartz HP	Si O2	1.215	Import(PDF-4 Minerals 2024)	152 : P3121	01-083-0542
Periclase	Mg O	3.128	Import(PDF-4 Minerals 2024)	225 : Fm-3m	00-001-1235

Phase Data View

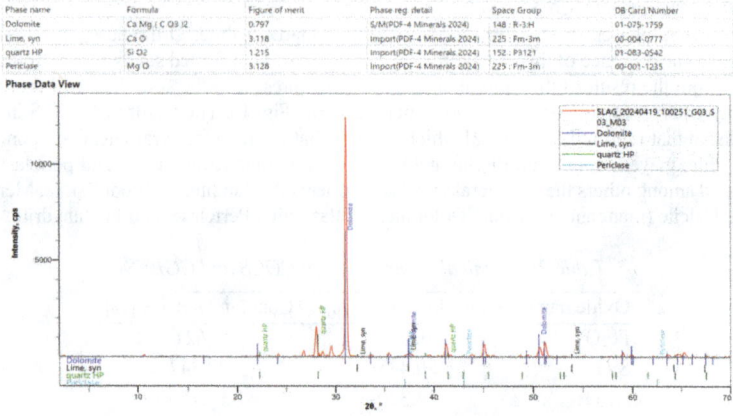

B: XRD for GGBFS

Fig. 1: X-ray Diffraction pattern for COCS and GGBFS

Total Petroleum Hydrocarbon (TPH)
According to the findings, the crude oil contamination soil samples of the Baranyowa-Dere area had an average percentage of TPH of 14.1 (Table 3) which was similar to the result obtained by [18] in the same region. This led to the decrease in pore spaces with an increase in the bulk density of the raw soil sample.

Table 3. Total Petroleum Hydrocarbon (TPH) Results

S/N	Sample Code	TPH [%]	AVG [%]
1	BDR1	14.34	
2	BDR2	14.87	14.1±1.1
3	BDR3	13.11	

Composition of Sodium Silicate Solution
Table 4 displays the result of the composition of sodium silicate (Na_2SiO_3) solution used in combination with sodium hydroxide (NaOH) as an activator. According to the results, the average percentage of silica (SiO_2) was 33.67%, Sodium oxide (Na_2O) was 15.92%, and total solid was 49.59%. The value related

Table 4: Chemical composition of Sodium silicate solution

Chemical composition	Results
Specific Gravity at 25 °C	1.416
Silica modulus (SiO_2/Na_2O)	2.11
Na_2O [%]	15.92
SiO_2 [%]	33.67
Total Solids [%]	49.59
Water Content [%]	50.41

Advances in Cement and Concrete
Materials Research Proceedings 51 (2025) 156-166

Materials Research Forum LLC
https://doi.org/10.21741/9781644903537-18

Geotechnical Properties of Natural COCS

The index properties shown in Table 5 demonstrated that the soil is dark grey, well-graded, fine-graded soil without inorganic clay and classed as an A-3 and ML based on the AASHTO and USCS classification system. Raw COCS was a non-plastic soil therefore it has a liquid limit of 0%, plastic limits of 0%, and plasticity indices of 0%. The grain size distribution of the natural soil is presented in Fig. 2. The Coefficient of Uniformity (C_u) and Coefficient of Curvature (C_c) of the sample are 3.0 and 1.2, respectively because D_{10} is zero according to the graph in Fig. 2.

Table 5: Properties of natural COCS Sample

Properties	Description value
Percentage passing sieve No 200	4.40
Coefficient of Curvature	1.2
Coefficient of Uniformity	3
Atterberg	Non-plastic soil
Unified Soil Classification	SW
AASHTO classification	A-3
Colour	Dark Grey
Specific Gravity	2.60
OMC_{BC} [%]	9.20
OMC_{WAS} [%]	8.80
MDD_{BS} [Mg/m^3]	1.89
MDD_{WAS} [Mg/m^3]	1.90
CBR_{BS} (Unsoaked) [%]	11.02
CBR_{BS} (after 48 hrs soaking) [%]	12.53
CBR_{WAS} (unsoaked) [%]	11.52
CBR_{WAS} (after 48 hrs soaked) [%]	14.03

NB: Subscript BC is British Proctor, while WAS is West African Standard

Fig. 2: Particle size distribution graph of raw COCS

Advances in Cement and Concrete   Materials Research Forum LLC
Materials Research Proceedings 51 (2025) 156-166   https://doi.org/10.21741/9781644903537-18

The percentage passing sieve No. 200 is 4.40% which shows that the sample contained less fine particles. Since the percentage of fine is less than 35%, according to [14], the soil is suitable for highway pavement construction because it will not be highly susceptible to swelling when in contact with water. The crude oil-contaminated soil (COCS) sample had a specific gravity of 2.60 while that of GGBFS is 2.89 g/cm^3. A good subgrade soil material should have a specific gravity in the range of 2.50 to 2.75 [15]. Therefore, the soil sample (COCS) has fallen within the acceptable specification. According to [16], the shear strength parameters (cohesion and angle of shearing resistance) would increase with a rise in specific gravity. Therefore, the result obtained indicates that the soil is suitable for use in the building of civil engineering structures, including retaining walls, buildings, bridges, and airfields.

Interaction of Activated GGBFS materials on OMC$_{WAS}$ and MDD$_{WAS}$
The Optimum Moisture Content (OMC) decreased with increase in the percentage of GGBFS and Activator when mixed with COCS, from 8.7 to 6.2% which is attributed to the decreased quantity of free soil, therefore smaller surface area needs less water. However, the maximum dry density (MDD) kept increasing from 1.9 to 2.08 Mg/cm^3 (Fig. 3a and 3b) which is attributed to the relatively higher specific gravity of GGBFS than that of the soil.

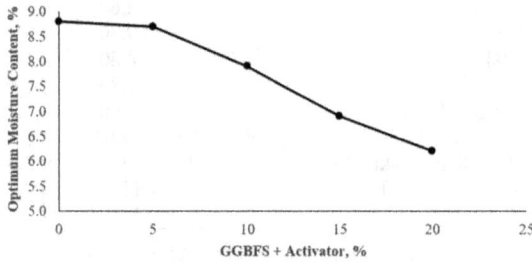

Fig. 3a: Effect of GGBFS and activator on the OMC$_{WAS}$ of the COCS sample

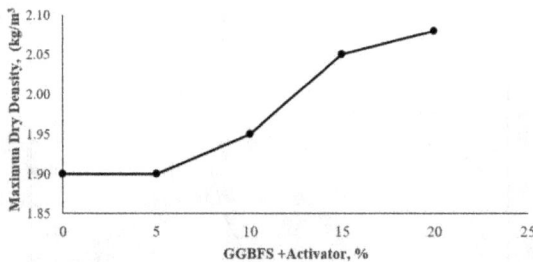

Fig. 3b: Effect of GGBFS and activator on the MDD$_{WAS}$ of the COCS sample

Advances in Cement and Concrete   Materials Research Forum LLC
Materials Research Proceedings 51 (2025) 156-166   https://doi.org/10.21741/9781644903537-18

Interaction of Activated GGBFS Materials on CBR$_{WAS}$

The California Bearing Ratio (CBR) test results for the soaked and un-soaked using West African Standard (WAS) compaction on soil samples are presented in Fig. 4. The values ranged from 19.04 to 52.57% and 13.98 to 39.38% for both soaked and unsoaked samples. Following [17], the recommended values to be used for sub-grade, sub-base, and base course materials should be $\geq$10%, $\geq$30%, and $\geq$80%, respectively. The (CBR) test for the Soaked soil samples shows that the COCS without additives can only serve as an excellent subgrade material. However, when additives GGBSF with activators were added, both soaked and unsoaked CBR kept increasing. The increase in CBR is due to the gradual development of a cementitious compound between the GGBFS and the Activator in the COCS sample [18].

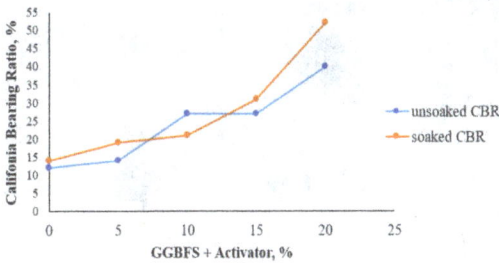

Fig. 4: Effect of GGBFS and activator on the CBR$_{WAS}$ of the COCS sample

Microstructural analysis
Microstructural and elemental study on COCS particles was carried out using Scanning Electron Microscope (SEM) coupled dispersive X-ray spectroscopy (EDS) for the analysis. The microstructures of COCS particles were roughly separate, loose and next to each other. From the sand micrograph in Fig. 5(a), it appeared that the connection between particles was not sticky.
Adding GGBFS and activator leads to a connection with the COCS atoms, thereby sticking composite atoms to each other. In the COCS stabilised with GGBFS and activator, additive particles fill the existing voids among the composite particles along with creating cementitious products. See Fig. 5(b-e).

The EDS analysis was performed to examine either chemical components or alterations of component amounts and their chemical composition. The element line spectrum results are shown in Fig. 6(a-e), with four different mixtures of COCS with GGBFS and activator composite. The element mapping results show the existence of Ca, Fe, Al, Si, Mg, Zr, Na and potassium (K) atoms. However, it was observed that the component of Si which is the greatest component reduced in proportion (70.85% to 40.76%) with an increase in the stabilized agent (GGBFS with activator). The mechanism of reaction in treated cases is driven by the ability of highly alkaline solutes, including NaOH, to dissolve amorphous silica and alumina source materials (mainly from GGBFS and possibly from the colloidal fraction of COCS into the matrix [19]. In such a condition, a reaction that chemically integrates minerals through alkaline activation may form a new structure [20-21]. The formation of this new structure may be responsible for the reduction of Si treated soil.

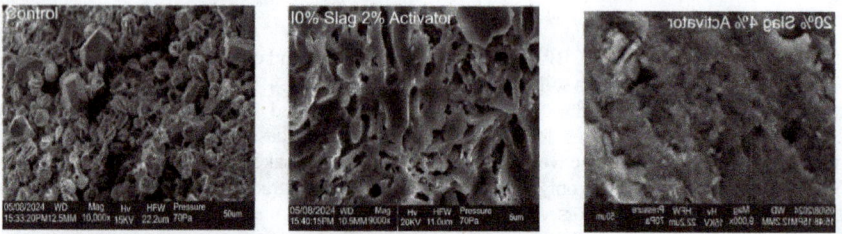

(a) 0%GGBFS 0% Activator   (b) 10%GGBFS 2% Activator   (c) 20%GGBFS 4% Activator

Fig. 5: SEM micrograph comparison of Soil sample

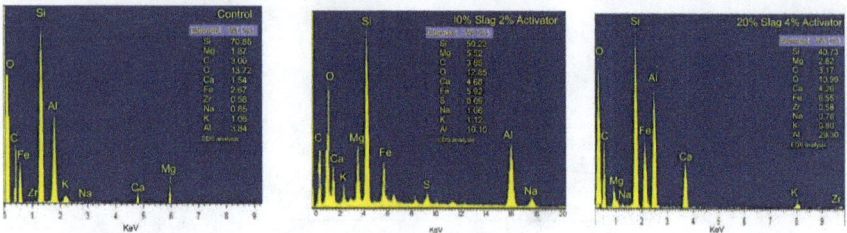

(a) 0%GGBFS 0% Activator   (b) 10%GGBFS 2% Activator   (c) 20%GGBFS 4% Activator

Fig. 6: EDS analysis comparison of Soil sample

Conclusions

The test's and analysis's findings led to the following conclusion.

i. The COCS sample belongs to A-3 well-graded sandy soil (good) under AASHTO classification which can only be suitable for filling and subgrade in highway construction.

ii. The interaction between COCS - GGBFS – Activator (NaOH+Na$_2$SiO$_3$) revealed that there is an increase in the strength by 78.93% and 77.38% at West African standard compaction (soaked and unsoaked) respectively.

iii. The percentage of GGBFS with activator used contributed to the strength properties of the COCS. Although we optimized the combination of the percentage of GGBFS (5%-20%) and activator (1%-4%), the bearing load of the soil measured by California Bearing Ratio (CBR) generally increased beyond expectation for all the mix ratio under CBR$_{soaked}$ and CBR$_{unsoaked}$ condition when compared to the CBR$_{soaked}$ and CBR$_{unsoaked}$ of unstabilized COCS.

iv. Activated GGBFS is a suitable binding material which is good for soil stabilization and has displayed potential to replace ordinary Portland Cement in its action.

References

[1] R.J. Oluremi, M.O. Osuolale, Oil Contaminated Soil as Potential Applicable Material in Civil Engineering Construction. *Journal of Environment and Earth Science*, 2014; 4(10): 87-99.

[2] N.J. Meegoda, N. Chen, D.S. Gunasekera, and P. Pederson, Compaction Characteristics of Contaminated Soils-reuse as a Road Base Material. *Geotechnical Special Publication*, 1998; 195-209.

[3] L.J. Provis, A.S. Bernal, Geopolymers and Related Alkali-Activated Materials. *Annual Review of Materials Research*, 2014; 44:299-327. https://doi.org/10.1617/s11527-013-0211-5.

Advances in Cement and Concrete
Materials Research Proceedings 51 (2025) 156-166

Materials Research Forum LLC
https://doi.org/10.21741/9781644903537-18

[4] K.L. Turner, G.F. Collins, Carbon Dioxide Equivalent (CO2-e) Emissions: A Comparison Between Geopolymer and OPC Cement Concrete. *Construction and Building Materials*, 2013; 43: 125-130. https://doi.org/10.1016/j.conbuildmat.2013.01.023

[5] M.D. Jean-Baptiste, F. Weipeng, L. Yanro, M. Lixin, D. Zhijun and Y. Jianqiao, Synthesis and Characterization of Alkali-Activated Loess and its Application as Protective Coating. *Construction and Building Materials,* 2021; 282: 122631. https://doi.org/10.1016/j.conbuildmat.2021.122631.

[6] T. Bakharev, J.G. Sanjayan andY.B. Cheng, Alkali activation of Australian slag cements. Cement and Concrete Research, 1999; 29: 113–120. https://doi.org/10.1016/S0008-8846(98)00170-7.

[7] Y. Li, Y. Sun, Preliminary study on combined-alkali–slag paste materials. *Cement and Concrete Research*, 2000; 30: 963–966. https://doi.org/10.1016/S0008-8846(00)00269-6.

[8] A. Adam, Strength and Durability Properties of Alkali Activated Slag and Fly Ash-Based Geopolymer Concrete. Ph.D. Thesis, RMIT University, Melbourne, Australia, 2009. (not published).

[9] J.J. Chang, A study on the setting characteristics of sodium silicate-activated slag pastes. *Cement and Concrete Research*, 2003; 33: 1005–1011. doi.org/10.1016/S0008-8846(02)01096-7.

[10] F.J. Martin, H.C. Doyne, Laterite and lateritic soil in Sierra Leone, II. *The Journal of Agricultural Science, 2009;* 20 (1): 135 – 143. https://doi.org/10.1017/S0021859600088675.

[11] B. Çubukçuoğlu, Use of Steel Industry By-products in Sustainable Civil Engineering Applications. *Proceeding from 8th International Conference on Environment Pollution and Prevention (ICEPP), Near East University, Nicosia, Northern Cyprus.*, 2020.

[12] J.R. Oluremi, W.H.A. Elsaigh, Characterization of coal fines as potential supplementary cementitious additive for remediation of non-polar substance contaminated soil. Materials Today: *Proceedings Volume 86*, 2023; 24–31. https://doi.org/10.1016/j.matpr.2023.02.055.

[13] Z.I. Yildirim, M. Prezzi, Chemical, Mineralogical, and Morphological Properties of Steel Slag. Advances in Civil Engineering, Article ID 463638, 13 pages, 2011.

[14] Federal Ministry of Works Highway Manual Part 1. Volume I: Geometric Design and Volume III: Pavements and Materials Design 2013 (Federal Republic of Nigeria)

[15] G.A. Oluyinka, C.O. Olubunmi, Geotechnical Properties of Lateritic Soil as Subgrade and Base Material for Road Construction in Abeokuta, Southwest Nigeria. *International Journal of Advanced Geosciences*, 2018; 6(1):78-82. https://doi:10.14419/ijag.v6i1.8952.

[16] S.T. Roy, G. Dass, Statistical models for the prediction of shear strength parameters at Sirsa, *India. International Journal of Civil and Structural Engineering*, 2014; 4(4): 483-498. https://doi:10.6088/ijcser.201404040002.

[17] J.S. Van Deventer, J.L. Provis, P. Duxson, and G.C Lukey, Reaction mechanisms in the geopolymeric conversion of inorganic waste to useful products. *Journal of Hazard and Materials*, 2007; 139(3): 506–513. https://doi.org/10.1016/j.jhazmat.2006.02.044.

[18] J. Dayalan, Comparative Study on Stabilization of Soil with Ground Granulated Blast Furnace Slag (GGBS) and Fly Ash. International Research Journal of Engineering and Technology, 2016; 3(5): 2198-2204.

[19] W.O. Ajagbe, O.S. Omokehinde, G.A. Alade, and O.A. Agbede, Effect of Crude Oil Contaminated Sand on the Engineering Properties of Concrete. *Construction and Building Materials*, 2012; 26(1): 9–12. https://doi:10.1016/j.conbuildmat.2011.06.028.

[20] P. Shahram, B.K.H. Bujang, A. Afshin and H.F. Mohammad, Model Study of Alkali-Activated Waste Binder for Soil Stabilization. *International Journal of Geosynthetics and Ground Engineering*, 2016; 2:35. https://doi 10.1007/s40891-016-0075-1.

[21] H. Xu, J.S. Van Deventer, The Geopolymerisation of Aluminosilicate Minerals. International Journal of Minerals Process, 2000; 59(3): 247–266. https://doi:10. 1016/S0301-7516(99)00074-5.

Keyword Index

About the Editors

RAHEEM Akeem Ayinde is a Professor of Building from Ladoke Akintola University of Technology (LAUTECH), Ogbomoso, Nigeria. He holds a Ph.D. in Building Structures from Obafemi Awolowo University Ile-Ife, Nigeria. He is presently the Head of Building Department (2021- till date). As an active researcher and project leader, he has led some funded researches and has more than eighty (80) articles in reputable journals locally and internationally, to his credit. He has successfully supervised Six (6) PhD theses, Twenty (20) M. Tech. dissertations and over 500 B.Tech. Students' projects while many are on-going. His current research interest is in incorporation of agriculture residues (such as rice husk ash, wood ash, corn cob ash, etc,) in cement and concrete.

Prof. Raheem has served as a reviewer to notable local and international journals such as USEP, Journal of Research Information in Civil Engineering (RICE), University of Ilorin, Nigeria; Construction Research Journal, Department of Building, University of Lagos, Nigeria; Journal of Construction and Building Materials (Elsevier), United Kingdom; Sustainability – Open Access Journal, Switzerland, etc. He has served as Guest Editor for Materials Today Proceedings, Volume 86, 2022. He is currently the Editor in Chief, LAUTECH Journal of Civil and Environmental Studies (LAUJOCES), Nigeria. He is the Team Leader, Cement and Concrete (CEMCON) Research Group. LAUTECH, Ogbomoso, Nigeria.

Prof Bolanle Deborah Ikotun is an Associate Professor at the University of South Africa (UNISA) in the Department of Civil and Environmental Engineering and Building Science. With over 20 years of experience in academia, her research focuses on sustainable and eco-friendly construction materials. She has studied extensively on concrete material characterisation, concrete mechanical and durability properties testing and concrete quality optimisation techniques by exploring the use of industrial and agricultural wastes. Driven by the vision of incorporating sustainable waste materials into construction materials engineering, she has addressed critical issues like carbon emissions and resource depletion in the construction industry. She has investigated various waste materials, including fly ash, steel slag, ferrochrome slag, rubber crumbs, glass wastes, plastic wastes, wood ash, palm kernel shell, wastewater sludge and paper sludge. She has produced over 70 peer-reviewed publications. She has presented at national and international conferences and collaborated across multidisciplinary fields. She has served as a reviewer for various reputable journals in the field of construction materials.

Her research activities have earned her international and local recognition including the 2010 Developing Researcher Award and the 2016 Resilience in Research Award at UNISA. Others include the Best Paper Award (International Conference of Manufacturing Engineering and Engineering Management,4-6 July 2018, London, U.K) and Sustainable Concrete Ambassador Award (1st International Conference on Advances in Cement and Concrete Research, 15-18 November 2022, LAUTECH, Nigeria). Her research on sustainable materials for concrete production has also been funded and supported by Volkswagenstiftung, managed by Bundesanstalt für Materialforschung und -prüfung (BAM), Germany (2015, 2016), NRF through the Knowledge, Interchange and Collaboration Programme (KIC) (2016, 2017, 2019, 2023) and Women in Research support grant for project (2018-2021). With the aim of transferring skills and promoting sustainability, she has supervised/co-supervised over 20 postgraduate students to completion.

She has served as a Guest Editor for "Multidisciplinary Digital Publishing Institute (MDPI) (Minerals, IF: 2.5)" in 2022, and a Managing Guest Editor for "Materials Today: Proceedings", Volume 86, in 2022.She has served as a Chair of the Research sub-committee at UJ, Civil Engineering Technology, Doornfontein Advisory board (2021-2023). Chair of the Departmental Research and Innovation Committee (DRIC) (2019-to date); Deputy Chair of the School of Engineering Research and Innovation Committee (2021-2023); the Chair of the School of Engineering Research and Graduate Studies Committee (2024-to date) and the head of the Sustainable Green Concrete Research group (2020-to date).